Jesus Christ the Solar Harvest Deity

William C. Henry Sr.

Disclaimer

46 St. Books
Published by 46 St. Books
46 St. Books USA Inc.,
Philadelphia, PA 19144, USA

Copyright © William C. Henry Sr., 2014

All rights reserved under International and Pan-American Copyright Conventions.

ISBN 13# 978-0-9916520-7-5   ISBN 10# 099165207X

Henry Sr., William.

PRINTED IN THE UNITED STATES OF AMERICA
Jesus Christ: The Solar Harvest Deity

PUBLISHER'S NOTE
This book is a work of non-fictional based on archival scriptures, texts, and documents. All images used are either owned, work for hire, acquired works, use with permission, and from the Library of Congress.

BOOKS ARE AVAILABLE AT QUANTITY DISCOUNTS WHEN USED TO PROMOTE PRODUCTS OR SERVICES. FOR INFORMATION PLEASE WRITE 46 ST. BOOKS, 55 WEST PENN ST, PHILADELPHIA, PA 19144 OR VISIT OUR WEBSITE AT
WWW.46STBOOKS.COM

Dedication

This is the first project that I dedicate to myself. The current period of evolving that I have passed through, has brought intense intellectual and spiritual super-standing. I say SUPER-standing, as we typically only UNDERstand, which continually lowers the bars of intellect.

I was told by sages that my path would cross the minds of the enlightened and the lesser so. The easy path would be to put my super-standing in my pocket, and thereby enjoy life unencumbered by ancient systems of enslavement. That was not the thing that my soul dictated that I do, as no one can be free, unless we all are free!

IPIWALUT

Table of Contents

The Sacred Template - page 9

Matthew and the Solar Christ - page 11

The Arabic Gospel of the Infancy of the Savior - page 87

Luke Decoded - page 97

The Disaster-Miracle Path of Jesus Christ - page 111

Jesus Christ: Deity of the Harvest - page 115

The Perihelion Miracle of Fatima - page 119

Anti-Masonic Prophecies and Miracles of Our Lady of Good Success - page 125

Conclusions - page 131

The Sacred Template

When I use the term template, it is in reference to the common tone and timbre used by Biblical, Apocryphal, and Sacred authors. This sacred template of understanding shows the authors common experiences and similar theology systems. The words are often rigid in their use, and other times flexible with variable meanings. The Prophets of the Old Testament were of the Watchman class, who were astronomers and priests combined, as they were Prophesying the course of the stars and their interactions with the Earth. This was an all important job that developed out of continual disaster Passovers, and these scientists priests were celebrated when they were accurate as saviors of mankind. The terms below will give you an excellent understanding of how to look at one at these so long misunderstood documents. The Watchmen of the Old Testament were the Goodmen of the New Testament, and were documenting the orbital path of Venus as she caused continual destruction in antiquity.

TEMPLATE

Angels: Planets, Moons, Stars, and Zodiac Bindings
Fallen Angels: Meteorites

Demons: Meteors, Comets, Asteroids, Planets, Stars, and Zodiac Bindings

Chariot: Circuit, Orbit - ex: Parabolic Trajectory

Horses: Meteor/Comet Trail

Sin: When a planet transgresses the Commandment of its orbit

Lord of Hosts: Sun, Venus, Jupiter, Saturn, etc.

Mansions/Court/Chamber: Windows, Pathways, Portals, Rooms, Gates, and Columns

Hell: Hyades constellation is a V-shaped cluster of stars in the Taurus constellation
Venus-Earth Synod Cycle 584 days - 243 years (Pentagram symbol) - 3 and 14 day cycles
below the ecliptic (Hell) plane out of sight and is reborn from the Evening Star as the Morning Star.

Passover: Planetary Passover

Venus: Satan, Lucifer, Quetzalcoatl, Kukulkan, Ishtar, Gukumatz, and over 40 more

Fixed: Circumpolar Orbit

Matthew & the Solar Christ

Chapter 1

*16 And Jacob begat **Joseph** the husband of **Mary**, of whom was born **Jesus**, who is **called Christ**.*

*17 So all the generations from **Abraham to David** are **fourteen generations**; and from **David** until the carrying away into **Babylon** are **fourteen generations**; and from the carrying away **into Babylon** unto **Christ are fourteen generations.***

A modern generation is typically considered to be twenty five years. A Biblical generation is considered to be anywhere from twenty five to one-hundred years. The figure that I have settled on is the old familiar fifty! Some researchers use 70 or 100 as the generation, but the timeline of peoples and events usually are skewered with those numbers. Fifty is one of the most revered numbers in the Bible being linked to the Sumerian Lord Enlil (El, Elul, Toro El, etc.) as 50 was his Sacred number. I will show below how 50 as a Biblical generation relates much better to people within that timeline.

(50) x 14 = 700 x 3 gen = 2100 yrs.
(70) x 14 = 980 x 3 gen = 2940yrs.
(100 x 14) = 1400 x 3 gen = 4200 yrs.
Abraham to Moses - 2100 BCE to 1400 BCE
Moses/David to Babylon - 1400 BCE to 700 BCE
Babylon to Christ 700 BCE to 0 CE

There is 1 complete Zodiac Age from Abraham to Christ 2160 years!

*18 Now the birth of Jesus Christ was on this **wise**: When as his mother **Mary** was **espoused** to **Joseph**, before they came together, she was **found** with **child** of the **Holy Ghost.***

The Son of God born of a virginal birth is a common tale that predates the Christian version by 2,000 years or more, as in the virginal birth of Horus in Egypt, as well as Krishna in India, and several dozen others.

Baal, Melqart, Adonis, Eshmun, Tammuz, Ra, Osiris, Dionysus, Quetzalcoatl (Venus), Innana (Venus), Ishtar (Venus), Persephone, Bari, Horus (Hours), Mithras, Attis, Adonis, Apollo, and others!

*Notice the all pervading fear and respect for Venus

*19 Then Joseph her **husband**, being a **just man**, and not willing to make her a **publick** example, was minded to put her away **privily**.*

*20 But while he **thought** on these **things**, behold, the **angel** of the **Lord** appeared unto him in a **dream**, saying, Joseph, thou **son of David**, fear not to take unto thee Mary thy **wife**: for that which is **conceived** in her is of the Holy Ghost.*

The common prophetic dream.

*21 And she shall bring forth a **son**, and thou shalt call his name **JESUS**: for he shall **save** his people from their **sins**.*

Translation: Jesus was born to teach people of the disasters to come from the planets sinning in their orbits.

*22 Now all this was done, that it might be fulfilled which was spoken of the Lord by the **prophet**, saying,*

*23 Behold, a **virgin** shall be with child, and shall bring forth a son, and they shall call his name **Emmanuel**, which being interpreted is, God with us.*

The statement by Isaiah that prophesizes the virginal birth of Emmanuel is interpreted as Jesus Christ being the fulfillment of the prophecy. Using the understanding that I have shown in regard to Isaiah being a Watchman (astronomer priest), we see the prophecy as relating to a heavenly body like Venus, and not that of the Son of God, but the Sun of God!

Isaiah {7:14} Therefore the Lord himself shall give you a sign; Behold, a virgin shall conceive, and bear a son, and shall call his name Immanuel.

24 Then Joseph being raised **from sleep did as the angel of the Lord had** bidden **him, and took unto him his wife:**

*25 And **knew** her **not** till she had brought forth her **firstborn son**: and he called his name **JESUS.***

CHAPTER 2

NOW when Jesus was born in **Bethlehem** of **Judæa** in the days of **Herod the king**, behold, there came wise men from the *east* to *Jerusalem*,

This long debated statement of clear astrotheology shows the 3 Wise Men (Alnilam, Alnitak, Mintaka) as the stars in the Belt of Orion constellation, who is represented as Herod (Orion) in this cosmic scenario. This birth in the East relates to the comet/Venus appearing in the sky in conjunction with the stars known as the 3 Wise Men.

2 Saying, Where is **he** that is born **King of the Jews?** for we have seen **his star** in the **east**, and are **come to worship** him.

The comet/Venus that was worshipped in the Old Testament in various embodiments, was held as the God or King of the Jews, because it took a circumpolar orbit over their Holy Land. His star in the East was Venus, as she is known as the Morning and Evening Star!

East Star- Eastern Star - Easter - Venus!

3 When Herod the king had heard *these things*, he was **troubled**, and **all Jerusalem with him.**

The king and all of Jerusalem are worried as the appearance of this body in the heavens brings fear of destructions.

4 And when he had **gathered** all the **chief priests** and **scribes** of the **people** together, he demanded of them where **Christ should be born.**

The chief priests and scribes would be used as astronomers and astrologers, in order to divinate the entrance of Venus into our parabolic (parabala, parable) orbit.

5 And they said **unto him**, In **Bethlehem of Judæa**: for thus it is **written by the prophet,**

The alignment will occur over Bethlehem and make landfall there.

6 And thou Bethlehem, *in* the land of Juda, art not the least among the **princes** of Juda: for out of thee shall come a

Governor, that shall rule my people Israel.

One is born to be Governor and rule the people of Israel! Jesus is clearly meant to be a local deity for the Jewish populace.

7 Then Herod, when he had **privily** called the **wise men**, inquired of them diligently **what time the star appeared.**

The timing was crucial for astrological divination, as we are constantly being connected to the stars in understanding..

8 And he **sent them to Bethlehem**, and said, Go and **search diligently** for the young child; and when ye have found *him*, **bring me word again, that I may come and worship him also.**

9 When they had heard the king, they departed; and, **lo, the star**, which they saw in the *east*, went before them, till it came and **stood over where the young child was**.

Venus takes a circumpolar orbit over Bethlehem in alignment with the 3 Wise Men, as the ejected material falls to the Earth.

10 When they **saw the star**, they **rejoiced** with **exceeding great joy**.

This is clear and present star and planetary worship as divinity!

11 And when they were **come into the house**, they saw the **young child with Mary** his mother, and fell down, and **worshipped** him: and when they had opened their treasures, they presented unto him **gifts; gold, and frankincense, and myrrh.**

Houses (windows, pathways, mansions, courts, etc.) are clear astronomical terms, as the lessons and theology was taught through the stars.

12 And being warned of **God** in a **dream** that they should not return to **Herod**, they **departed** into their **own country** another way.

13 And when they were **departed**, behold, the **angel** of the Lord **appeareth** to Joseph in a **dream**, saying, **Arise**, and take the young child and his mother, and **flee** into **Egypt**, and be thou there until I bring thee word: for Herod will **seek** the young child to **destroy** him.

The orbital path takes Venus over the land of Egypt ahead of the Orion/Herod constellation.

14 When he **arose**, he took the young child and his mother by **night**, and **departed into Egypt**:

We have another nocturnal event for Jesus Christ. We typically see Jesus doing his ministry and work during the day, but nearly 100% of events noted are from the evening on, which underscores a parallel as the Evening Star of Venus!

15 And was there until the death of Herod: that it might be fulfilled which was spoken of the Lord by the prophet, saying, Out of Egypt have I called my son.

The basis for our calendar system of dating seems to lineup more with the death of Herod (Orion) as opposed to the birth of Christ.

16 Then Herod, when he saw that he was **mocked** of the **wise men**, was exceeding **wroth**, and **sent forth**, and **slew** all the **children** that were in **Bethlehem**, and in all the **coasts** thereof, from **two years old and under**, according to the **time** which he had **diligently** inquired of the **wise men**.

We have the unproven Slaughter of the Innocent. An event similar to the curse suffered by the selectivity of the Exodus disasters with the death of the first born. The Passover event causes the deaths of the children of Bethlehem as the 3 Wise Men signaled the time and arrival.

17 Then was fulfilled that which was spoken by **Jeremy** the prophet, saying,

Jeremy = Jeremiah

18 In **Rama** was there a voice heard, lamentation, and weeping, and great mourning, Rachel weeping *for* her children, and would not be comforted, because they are not.

Rama/Ramah - an elevated spot
Rama - Ramadan - Brama - Abram

19 But when **Herod** was **dead**, behold, an **angel** of the Lord **appeareth** in a **dream** to **Joseph** in **Egypt**, 20 Saying, **Arise**, and take the **young child** and his **mother**, and go into the **land** of **Israel**: for they are **dead** which **sought** the young **child's** life.

We have an orbital path change from Egypt (DMP) to Israel (DMP).

21 And he arose, and took the young child and his mother, and came into the land of Israel.

22 But when he heard that **Archelaus** did reign in Judæa in the **room** of his father Herod, he was afraid to go thither: notwithstanding, being warned of God in a **dream**, he turned aside into the parts of **Galilee**:

Orbital path correction to Galilee versus Judea.

Room: houses, mansions, pathways, are astronomy terms

23 And he came and dwelt in a city called **Nazareth**: that it might be fulfilled which was spoken by the prophets, He shall be called a **Nazarene**.

A circumpolar orbit over Nazareth is taken, and will become known as the Nazarene!

*The Nazirite (Nazi - Ashkenazi) vow could run for a day or a lifetime, depending upon the persons choice.
Chapter 3

2 And saying, **Repent** ye: for the **kingdom of heaven is at hand**.

The kingdom of heaven is at hand, as Venus is set to bring judgment and destruction to the sinners.

3 For **this** is **he** that was **spoken** of by the **prophet Esaias**, saying, The **voice** of one **crying** in the **wilderness**, **Prepare** ye the **way of the Lord**, make his **paths straight**.

The Watchman Isaiah (Esaias) forecast (prophesized, predicted) the events that are coming to pass. I showed in Isaiah 10:28 to 10:32 where Isaiah as a Watchman, makes observations and predictions of where the orbital path of Venus will be. He notes the cities that are affected in this narrow swath that follows a straight path! The two passing planetary bodies makes a loud noise during the event.

Isaiah 40:3 The voice of him that crieth in the wilderness, Prepare ye the way of the LORD, make straight in the desert a highway for our God.

Isaiah is clearly telling the scientifically unenlightened masses, that they have to allow the path of Venus to pass straight by, and not to follow after it lest they perish

6 And were **baptized** of him in **Jordan, confessing** their **sins**.

The orbital Passover of Venus deposited minerals and contaminants in the Jordan River that the people were baptized in to expiate their sins. This baptismal probably caused more infections than cleansings.

7 But when he saw many of the **Pharisees** and **Sadducees** come to his **baptism**, he said unto them, O **generation** of **vipers**, who hath **warned** you to **flee** from the **wrath** to come?

I believe that the Pharisees and Sadducees made their own celestial observations, and did not believe that this was the Sun/Son of God that was to come. The Old Testament believers recognized God (Wheel In the Sky) as the destroyer of note during that period, and did not see Venus (Wheel In the Sky) as the feature noted in the heavens as she changed shaped and transfigured over time. The event was due to unfold over more northern regions of Galilee, and not over Jerusalem, the Holy Land of the Pharisees and Sadducees.

11 I indeed **baptize** you with **water** unto **repentance**: but he that **cometh after me is mightier than I,** whose **shoes** I am not **worthy** to bear: he shall baptize you with the **Holy Ghost**, and *with* **fire**:

The watery deluge from the skies will pale in comparison to what will occur when Venus (Holy Ghost/Host) baptizes the region with fire.

13 **Then** cometh **Jesus** from **Galilee** to **Jordan** unto **John**, to be **baptized** of him.

The orbital path takes Venus (Jesus) from Galilee to Jordan to make landfall in the Jordan River to be baptized.

16 And Jesus, when he was baptized, went up straightway out of the water: and, lo, the heavens were opened unto him, and he saw the Spirit of God descending like a dove, and lighting upon him:

The ejection disk from Venus (Jesus) heads back to the heavens as the sky parts from the Passover event with lightning flashing.

17 And lo a **voice** from **heaven**, saying, This is my beloved **Son**, in whom I am well **pleased**.

The New Testament recognition of Jesus as the Sun/Son of God. The all consistent Sun God theme.

CHAPTER 4

THEN was **Jesus** led up of the **Spirit** into the **wilderness** to be **tempted** of the **devil**.

Venus (Jesus) takes an orbital path away from Galilee.

2 And when he had **fasted forty days and forty nights**, he was afterward an hungered.

Venus (Jesus) is out of sight to the worshippers for the common forty day orbital period.

3 And when the **tempter** came to him, he said, If **thou** be the **Son of God, command** that these **stones** be made **bread**.

The stones would be the meteorites, and the bread was the manna (Exodus) that was deposited with each Passover event.

Neph = stone   aka   Nephilim = Stones from heaven

5 Then the devil taketh him up into the holy city, and setteth him on a pinnacle of the temple,

The holy city in the cosmic heavens.

6 And saith unto him, If thou be the **Son of God**, cast thyself **down**: for it is written, He shall **give** his **angels** charge concerning thee: and in *their* **hands** they shall **bear** thee up, **lest** at any time thou dash thy foot against a **stone**.

Venus (Jesus) is being challenged/tempted to show that she/he is the Sun/Son of God or she/he would be able to fall/cast down from the temple.

7 **Jesus** said unto him, It is **written** again, Thou **shalt** not **tempt** the Lord thy God.

8 Again, the **devil** taketh him up into an exceeding **high mountain**, and **sheweth** him all the **kingdoms** of the world, and the **glory** of them;

The events always take place on high mountain places, as this is where the ancient Gods lived in the understanding of the people.

12 Now when Jesus had heard that John was cast into prison, he departed into Galilee;

The orbital path of Venus (Jesus) takes it away towards Galilee. It always struck me as odd, why Jesus never went to get John out of prison?

13 And leaving Nazareth, he came and dwelt in Capernaum, which is upon the sea coast, in the borders of Zabulon and Nephthalim:

The orbital path moves from Nazareth to Capernaum by the Sea of Galilee near Zabulon and Nepthalim.

Nepthalim / Napthali = Lebanon area (Neph - kidneys - stones - meteorites!)

14 That it might be **fulfilled** which was spoken by **Esaias** the prophet, saying,

15 The **land of Zabulon**, and the **land of Nephthalim**, by the **way of the sea, beyond Jordan, Galilee of the Gentiles;**

Zabulon: Lebanon
Nephthalim: Lebanon/Syria

16 The people which ***sat in darkness saw great light; and to them which sat in the region and shadow of death light is sprung up.***

The planetary eclipse caused by the Passover of Venus brings darkness to one region and light to another.

17 From that time Jesus began to preach, and to say, **Repent: for the kingdom of heaven is at hand.**

The planetary Passover event is set to occur.

18 And Jesus, **walking by the sea of Galilee**, saw **two brethren**, **Simon** called **Peter**, and **Andrew** his brother, **casting** a net into the **sea**: for they were **fishers**.

19 And he saith unto them, **Follow me**, and I will make you **fishers of men**.

20 And they **straightway** left *their* nets, and **followed** him.

21 And going on from thence, he saw **other two brethren**, **James** *the son* of **Zebedee**, and **John his brother**, in a ship with Zebedee their father, **mending** their nets; and **he called them.**

22 And they **immediately** left the **ship** and their father, and **followed** him.

We see that the 12 Disciples are picked up along the route, as the 12 signs of the zodiac are disciples of the Sun and follow it through the heavens.

12 Apostles - Peter, James, John, Andrew, Phillip, Judas, Thomas, Matthew, James, Simon, Thaddeus, and Bartholomew

12 Zodiac Signs - Aquarius, Pisces, Gemini, Leo, Virgo, Aries, Scorpio, Sagittarius, Taurus, Capricorn, Cancer, and Libra

12 Tribes of Israel - Reuben, Simeon, Levi, Judah, Dan, Napthali, Gad, Asher, Issachar, Zebulun, Joseph, and Benjamin (Dinah was Virgo originally)

* This is why we have 12 to a dozen and many more

23 **And Jesus went about all Galilee**, teaching in their **synagogues**, and preaching the gospel of the kingdom, and **healing** all manner of **sickness** and all manner of **disease** among the **people**.

24 And his **fame** went throughout all **Syria**: and they brought unto him all sick people that were taken with divers diseases and **torments**, and those which were **possessed** with **devils**, and those which were **lunatick**, and those that had the **palsy**; and he healed them.
Syria plays heavily in the New Testament.

Lunatics: moon worshippers
Lughnasadh: pronounced "loo-ne-se " lunacy
CHAPTER 5

AND seeing the multitudes, he **went up into a mountain**: and when **he was set**, his **disciples came unto him**:

Evening event on a mountain (Evening Star - Venus).

18 For **verily** I say **unto** you, Till **heaven** and **earth pass**, one **jot** or one **tittle** shall in no **wise** pass from the **law**, till all be **fulfilled**.

This is another often misinterpreted verse. We have to remember that Jesus is speaking in terms the populace should be able to understand. Until the Passover of Venus (heaven) beyond the Earth, then the laws of planetary bodies will be fulfilled and not one dot or comma will change the parabola!

29 And if thy right eye offend thee, pluck it out, and cast it from thee: for it is profitable for thee that one of thy members should perish, and not that thy whole body should be cast into hell.

Actually read: And if your right eye is infected beyond repair, it should be surgically removed: as it is better to lose an appendage, than to die from an infection!

30 And if thy right hand offend thee, cut it off, and cast it from thee: for it is profitable for thee that one of thy members should perish, and not that thy whole body should be cast into hell.

Actually read: If your right hand is infected, it should be surgically removed: as it is better to lose an appendage, than to die from an infection!

The man that we have come to know as Jesus Christ shows high levels of medical understanding, and the statement is being understood as theology versus medical!

CHAPTER 6

TAKE **heed** that ye do **not** your **alms** before **men**, to be **seen** of them: **otherwise** ye have **no reward** of your Father which is **in heaven**.

The act of true giving is not one of recognition!

5 And when thou prayest, thou shalt not be as the ***hypocrites** are:* for they ***love to pray standing*** in the ***synagogues*** and in the ***corners of the streets***, that they may be seen of men. Verily I say unto you, They have their
reward.

We see that 2,000 years later that men still pray in this fashion in the Holy Land (Jewish, Islamic, and Christian).

7 But when ye pray, use not ***vain repetitions***, as the heathen *do:* for they think that they shall be heard for their much speaking.

Mankind still uses repetitious self-hypnotic prayers and chants on a daily basis.

9 After **this manner** therefore pray ye: Our **Father** which **art in heaven**, **Hallowed** be thy name.

Venus is recognized as the seat of heaven.

10 Thy **kingdom come**. Thy **will** be **done** in **earth**, as *it is* in **heaven**.

The will of Venus from the heavens will play out on the Earth when she arrives.

11 **Give** us **this** day our **daily bread**.

The carbogenous content that we know as manna, mahdu, or ambrosia, is being deposited with the Passover of the Morning Star Venus.

12 And **forgive** us our **debts**, as we **forgive** our **debtors**.

This is the common Passover tradition of the Hebrew people, as in the 49th year all debts, slave, and land ownership was relinquished.

13 And **lead** us **not** into **temptation**, but **deliver** us from **evil**: For **thine** is the **kingdom**, and the **power**, and the **glory**, for ever. **Amen**.

The prayer is for deliverance from the evil that would result from a close Passover. Amen, is the Egyptian Sun God aka Amen Ra. The term Amen is wholly Egyptian, and is not found anywhere else!

CHAPTER 8

*1. **WHEN he was come down from the mountain, great multitudes followed him.***

We have an orbital descent as the people follow Venus in adoration and fear.

2 And, behold, there came a **leper** and worshipped him, saying, Lord, if thou wilt, thou canst **make me clean**.

3 And Jesus put forth *his* hand, and **touched** him, saying, I will; be thou clean. And ***immediately*** his ***leprosy*** was ***cleansed***.

Leprosy is the most common ailment in the Bible and in antiquity. The leprosy is related to iron sulfide and is not curable, but the sufferer is able to live a long life after. The Dead Sea (Pale Horse) and the Sea of Galilee (White Horse) are two of the Disaster - Miracle (DMP) bodies of water in almost a straight orbital path, and both are associated with treatments for various maladies. Te earliest documented case of leprosy is found in a skeleton in India from 1400 BCE (Exodus disaster period).

5 And when **Jesus** was entered into **Capernaum**, there **came** unto him a **centurion**, beseeching him,

Jesus Christ of Capernaum (DMP), as 98% of his miracles occur in this region.

6 And saying, Lord, my servant lieth at home sick of the **palsy**, grievously tormented.

Bell's Palsy (partial paralysis of facial nerves - facial drooping)

*Bulbar Palsy (impairment of cranial nerves - difficulty swallowing, slurring of speech, difficulty using voice)
Cerebral Palsy (neural disease or radiation exposure?)*

Conjugate gaze palsy (ability to move the eyes - stroke)

Erb's Palsy (inability to move an arm - difficult labor) Spinal Muscular Atrophy (wasting palsy - genetic - sit frog legged, limp limbs)?

8 The **centurion** answered and said, Lord, I am not worthy that thou **shouldest** come **under** my **roof**: but speak the **word** only, and my **servant** shall be healed.

As Venus Passes Over the centurion's house, the cometary deposition filtered inside to help or heal the servant.

11 And **I say** unto you, That **many** shall come from the **east and west**, and shall **sit down** with **Abraham**, and **Isaac**, and **Jacob**, in the **kingdom of heaven**.

Jesus knows that people will make Pilgrimages to the Holy Land to worship Venus as the kingdom of heaven.

12 But the **children of the kingdom** shall be **cast** out into **outer darkness**: there shall be **weeping** and **gnashing** of **teeth**.

The Pilgrims will be cast into total darkness with the Passover of Venus, and there will be fear unparalleled.

16 When the **even** was come, they brought unto him many that were ***possessed*** with devils: and he cast out the **spirits** with *his* word, and **healed** all that were **sick**:

Evening Star Venus!

18 Now when Jesus saw great **multitudes** about him, he gave **commandment** to **depart** unto the **other side**.

The orbital path moves across the Sea of Galilee.

20 And Jesus saith unto him, The foxes have holes, and the birds of the air *have* nests; but the **Son of man** hath not where to **lay** *his* **head**.

Sun of Man!

22 But Jesus said unto him, Follow me; and let the dead bury their dead.
I have no idea!

Chapter 9

And he **entered into a ship**, and **passed over**, and came into **his own city.**

Venus moves over the Sea of Galilee toward Nazareth.

[2] And, **behold**, they brought to him a man **sick** of the **palsy**, **lying** on a **bed**: and Jesus seeing their **faith** said unto the **sick** of the palsy; Son, be of **good cheer**; thy **sins** be **forgiven** thee.

Did the invalid suffer from Cerebral Palsy? There is a common placement of sins with the suffering and diseases engendered from these Passover events.

[7] And he **arose**, and **departed** to his **house**.

There is a difference in belief about the feature in the sky as it moves to a different (windows, rooms, pathways, mansions, etc) house in the sky.

[8] But **when** the multitudes **saw** it, they **marvelled**, and **glorified God**, which had given such **power** unto **men**.

The uneducated populace believe this to be a mighty weapon from God that was given to men.

[12] **But when Jesus heard that, he said unto them, They that be whole need not a physician, but they that are sick.**

[13] **But go ye and learn what that meaneth, I will have mercy, and not sacrifice: for I am not come to call the righteous, but sinners to repentance.**

The righteous would have understanding as to their medical needs, and of the events that were to Passover. The class of doctors did not eat with the lower level people.

[15] And Jesus said unto them, Can the children of the **bridechamber** mourn, as long as the **bridegroom** is with them? but the **days will come**, when the bridegroom shall be **taken from them**, and then shall they **fast**.

Bride: Andromeda constellation
Bridegroom: Perseus constellation

Catholic nuns are wed to Jesus in a weird one sided ceremony

[20] And, behold, a **woman**, which was **diseased** with an issue of **blood twelve years**, came behind him, and **touched** the **hem** of his **garment**:

Blood diseases: Various anemia's, lymphomas, leukemia, coagulation and hemorrhagic conditions, and more.

[32] As they **went out**, behold, they brought to him a **dumb** man **possessed** with a **devil**.

Bulbar Palsy (impairment of cranial nerves - difficulty swallowing, slurring of speech, difficulty using voice)?

[34] But the **Pharisees** said, He **casteth** out **devils** through the **prince** of the **devils**.

The Pharisees clearly correlate these cures to the planet known also Lucifer (Venus) for these miraculous cures.

*The Vatican owns an astronomical observatory called Lucifer!

[35] And **Jesus** went about all the **cities** and **villages**, teaching in their **synagogues**, and **preaching** the gospel of the **kingdom**, and **healing** every **sickness** and every **disease** among the people.

There will be those that argue eloquently, and others in a more vehement manner. I believe that the man that we have come to know as Jesus Christ was of the Watchman/Goodman class, physician, astronomer, teacher and more. Jesus prophesized the disasters to come from the heavens above, and these events were anthropomorphized in a seemingly smooth synthesis with the real deeds of a man long after his death.

[36] But when he **saw** the **multitudes**, he was **moved** with **compassion** on them, because they **fainted**, and were **scattered** abroad, as **sheep** having no **shepherd**.

The Pilgrims faint and scatter (fainting sheep) as the Passover unfolds.

[37] Then saith he unto his **disciples**, The **harvest** truly is **plenteous**, but the **labourers** are **few**;

The growth cycles of man, plant, and animal, may have been sped up due to the disturbance in the thermal-electric tide.

[38] Pray ye therefore the **Lord of the harvest,** that he will send forth labourers into his harvest.

The Lord of the Harvest reference is a clear connection to the Greek God of the Harvest Pan! God of Nature, the Wild, Shepherds, Flocks, Goats, and of Mountain Wilds. A Christian temple in Capernaum is built on an old temple of Pan!!!

The first Christmas was celebrated in Rome at the Lupercal pagan shrine, with wine and child sacrifices!

Chapter 10

10 And when he had called unto him his **twelve disciples**, he gave them power against **unclean spirits**, to cast them out, and to **heal all manner** of **sickness** and all manner of **disease**.

The 12 Disciples represent the 12 Zodiac signs that orbit around the Sun, as Venus (Jesus) is bestowed the appellation the Sun of God. The 12 Tribes of Israel represent the 12 Zodiac bindings, and this is why the New Testament Christians constantly endeavor to equal or outdo the Old Testament's symbolism.

[5] These **twelve** Jesus **sent forth**, and **commanded** them, saying, Go not into the way of the **Gentiles**, and into any city of the **Samaritans** enter ye not:

The orbital path is to miss the lands of the Gentiles (DMP) and Samaritans (DMP).

[6] But go rather to **the lost sheep** of the **house** of **Israel**.

This Passover event is to play out in Jerusalem.

[8] **Heal** the **sick**, **cleanse** the **lepers**, **raise** the **dead**, **cast** out **devils**: freely ye have **received**, freely **give**.

The disciples/apostles are trained as physicians of the time versus theologians.

¹⁴ And whosoever shall not **receive** you, nor **hear** your **words**, when ye **depart** out of that **house** or **city**, **shake off** the **dust of your feet.**

Jesus is showing the disciples how to avoid contamination! The dirt and soil mixed with cometary debris would contain contaminants that needed to be washed from the bare feet/sandals and clothing when entering or leaving an affected region.

¹⁵ Verily I say unto you, It shall be **more tolerable** for the land of **Sodom** and **Gomorrha** in the day of **judgment**, than for that **city**.

When judgment comes to Passover, the lands of Sodom and Gomorrha will not be in the orbital path, but Jerusalem will be.

¹⁷ But **beware** of **men**: for they will **deliver** you up to the **councils**, and they will **scourge** you in their **synagogues**;

¹⁸ And ye shall be **brought** before **governors** and **kings** for my sake, for a **testimony** against **them** and the **Gentiles**.

Govern - Governor - Government

³⁴ Think not that I am come to send peace on earth: I came not to send peace, but a sword.

This statement shows the kinship in theology with Hindu and Mayan cultures. The Hindu Goddess Kali Ma (Venus), the Mayan God Quetzalcoatl (Venus), and Jesus Christ (Venus), are all pictured with swords (ejection disk, privy member, penis) protruding from their mouths to destroy and judge from the celestial heavens.

[38] And he that taketh not his cross, and followeth after me, is not worthy of me.

Prophecy or retro vision? The Christian concept of the Crucifixion is predated by a list of over thirty Solar and Celestial deities such as; Dionysus, Appollo, Horus, Attis, Mithras, Krishna, and many more!

Chapter 11

[7] And as they departed, Jesus began to say unto the multitudes concerning John, What went ye out into the wilderness to see? A reed shaken with the wind?

The astrological feature that John embodied was no longer there, as they wished to see the might of the tempest.

[9] But what went ye out for to see? A prophet? yea, I say unto you, and more than a prophet.

Prophet: Planet, star, zodiac binding

¹⁰ For this is he, of whom it is written, Behold, I send my messenger before thy face, which shall prepare thy way before thee.

The commandment (orbital path) was written that the event would be preceded by meteorites (messengers) before the full planetary Passover. Jesus is constantly trying to teach and warn of the type of event to come.

¹² And from the days of John the Baptist until now the kingdom of heaven suffereth violence, and the violent take it by force.

The celestial heavens have been in turmoil with planetary upheaval during the recent Jubilee event (49 yrs.) as it is supposed that John was only a couple of years older than the figure of Jesus.

²¹ Woe unto thee, Chorazin! woe unto thee, Bethsaida! for if the mighty works, which were done in you, had been done in Tyre and Sidon, they would have repented long ago in sackcloth and ashes.

Chorazin and Bethsaida have been decimated from the Passovers, and the inhabitants of Tyre and Sidon of Lebanon, are not so fast to give up their deities like Baal of Tyre, Baal of Carthage, and Baal Teshuva?

Chorazin and Bethsaida sit right in the Disaster-Miracle Path near the Sea of Galilee!

²² But I say unto you, It shall be **more tolerable** for **Tyre** and **Sidon** at the **day** of **judgment**, than for you.

Tyre and Sidon will be on the borders of the orbital path avoiding a direct hit on Judgment Day. JUDGMENT DAY has ALREADY Passed Over people! It is time for all of humanity to come to a new understanding, that we may stop the senseless genocide and other atrocities enacted in the name of Deism or Theism!

²³ And thou, **Capernaum**, which art **exalted** unto heaven, shalt be brought **down to hell**: for if the mighty **works**, which have been done in thee, had been done in **Sodom**, it would have remained until this day.

I previously felt that Capernaum exhibited scorching on the remaining ruined synagogues! The smaller portion of rock used is volcanic (black basalt) but the residue covers larger buildings and columns that are white calcareous stone. The works done in Sodom were of a FAR GREATER level than the event that destroys Capernaum.

²⁵ At that **time** Jesus **answered** and said, I thank thee, O **Father**, **Lord of heaven and earth**, because thou hast **hid** these **things** from the **wise and prudent,** and hast revealed them unto **babes**.

The wise and prudent Sadducees and Pharisees do not recognize the same astronomical signs in the skies, and disciples and followers are the babes.

[30] For my **yoke** is easy, and my **burden** is **light**.

Something about being yoked (bridal-bridle) bothers me. Animals are yoked together!

Chapter 12

12 At that time **Jesus** went on the **sabbath** day through the **corn**; and his **disciples** were an **hungred**, and began to **pluck** the **ears** of corn and to **eat**.

The Christian sabbath is on Sunday (Sun) as the event causes the corn field to be damaged (popped?).

Beth-dagon = Field of Corn/Fish!

[8] For the **Son of man** is **Lord** even of the **sabbath** day.

Venus as Sun/Son of man rules as Lord even on Sunday (Sun) as she shines brighter in the sky. The Jewish sabbath is on Saturday (Saturn's day).

[9] And when he was **departed** thence, he went into their **synagogue**:

Orbital path moves.

[10] And, **behold**, there was a **man** which had his **hand withered**. And they asked him, saying, Is it **lawful** to **heal** on the **sabbath** days? that they might **accuse** him.

I have shown in the Arabic Gospel of the Infancy of the Savior, where Jesus withers the hand of a teacher that strikes him, and then strikes him dead.

¹⁴ Then the Pharisees went out, and held a council against him, how they might destroy him.

The Pharisee tried to figure how to destroy the planetary Passover of Venus! NASA and many world governments are still hard at work with this type of scenario, as they believe in it as a possible eventually of our existence in a fluid universe.

²⁴ But when the Pharisees heard it, they said, This fellow doth not cast out devils, but by Beelzebub the prince of the devils.

The Pharisee recognize that it is Venus aka Beelzebub aka Lucifer this is casting demons (meteorites) out.

²⁵ And Jesus knew their thoughts, and said unto them, Every kingdom divided against itself is brought to desolation; and every city or house divided against itself shall not stand:

²⁶ And if Satan cast out Satan, he is divided against himself; how shall then his kingdom stand?

Jesus understood their astronomical protests, as he tries to show that the event will lessen as material is cast from Satan (Venus) during the Passover events, and will eventually run out of demons (meteorites) in its kingdom.

³⁸ Then certain of the scribes and of the Pharisees answered, saying, Master, we would see a sign from thee.

The scribes and the Pharisee wish that Venus (Jesus) would perform a Sign in the heavens.

³⁹ But he answered and said unto them, An evil and adulterous generation seeketh after a sign; and there shall no sign be given to it, but the sign of the prophet Jonas:

⁴⁰ For as Jonas was three days and three nights in the whale's belly; so shall the Son of man be three days and three nights in the heart of the earth.

The sign of Jonas/Jonah is total astrotheology and speaks of a 3-day orbital event, as the Whale constellation aka the Cetus (Beast/Sea Monster) constellation is featured prominently during the Age of Aries and Pisces.

⁴² The queen of the south shall rise up in the judgment with this generation, and shall condemn it: for she came from the uttermost parts of the earth to hear the wisdom of Solomon; and, behold, a greater than Solomon is here.

The queen of the South as Sheba is another zodiacal story related with theology. Sheba is the constellations of Andromeda and Cassiopeia, and Solomon is anthropomorphized as the Perseus constellation.

43 When the unclean spirit is gone out of a man, he walketh through dry places, seeking rest, and findeth none.

When the Earth is dried from the Passover then Venus (unclean spirit) will depart.

44 Then he saith, I will return into my house from whence I came out; and when he is come, he findeth it empty, swept, and garnished.

The house of the Sun!

45 Then goeth he, and taketh with himself seven other spirits more wicked than himself, and they enter in and dwell there: and the last state of that man is worse than the first. Even so shall it be also unto this wicked generation.

That wicked generation has passed, and many more have followed! The seven other spirits; Mercury, Mars, Saturn, Jupiter, Uranus, Neptune, and the Moon. Utter astrotheology espoused in Biblical context.

Chapter 13

13 The same **day** went Jesus **out** of the **house**, and **sat** by the **sea side**.

This speaks to a circumpolar orbit!

² And great **multitudes** were gathered together unto him, so that he **went into a ship**, and sat; and the whole multitude stood on the **shore**.

³ And he **spake** many things **unto** them in **parables**, saying, Behold, a sower went forth to sow;

Jesus spoke to the masses as did Enoch in parables. The teachings did not sink in with the masses of uneducated people. The easiest way to get the needed instructions across was through the use of recognizable and memorable tales. The prefix para has many meanings, including: alongside of, beside, near, resembling, beyond, apart from, and abnormal. The story is being told in a roundabout manner for the narrative to be understood!

PARA: at or to one side of, beside, side by side,
PARABLE: a simple story used to illustrate a moral or spiritual lesson.
PARABOLIC : of or like a parabola, expressed by or being a parable.
PARABOLIC TRAJECTORY : of, having the form of, or relating to a parabola motion in a parabolic *curve.*

First known use of the word parable was in 1669!

A parabolic trajectory is known as a Kepler Orbit, and is one of his Planetary Laws of Motion.

¹⁰ And the **disciples** came, and said unto him, Why **speakest** thou **unto** them in **parables**?

¹¹ He answered and said unto them, Because it is given unto you to know the mysteries of the kingdom of heaven, but to them it is not given.

The flock does not have the understanding of what is to come, and how to prepare themselves.

¹² For whosoever hath, to him shall be given, and he shall have more abundance: but whosoever hath not, from him shall be taken away even that he hath.

¹³ Therefore speak I to them in parables: because they seeing see not; and hearing they hear not, neither do they understand.

Jesus has to dummy down the information in order for the people to see, hear, and truly understand! The teaching of parables versus parabolics aka parabalas (Kepler Orbit) has led to mankind to not understand the lessons that Jesus imparted, due to improper understanding and re-teaching the misunderstanding as fact. If we use Nikolai Tesla and Neil deGrasse Tyson as examples we may be able to draw parallels to what Jesus was faced with. The mind of Nikolai Tesla spoke totally to him of numbers and theories. A person with qualities of this nature often are incapable of speaking to a lay person in terms that are easy for their understanding. The wonderful genius that flowed from his mouth would simply sound like a foreign language to the larger section of us. Enter the massive intellect of Dr. Tyson with an engaging personality and loquacious ability that I thoroughly enjoy. Dr. Tyson is capable of conveying complex astrophysics to the common folk with amazing ease and grace. The teachings that Jesus imparted were to get people out of the orbital path to avoid death and disease!

[14] And in **them** is **fulfilled** the **prophecy** of **Esaias**, which saith, By **hearing** ye shall **hear**, and shall **not understand;** and seeing ye **shall see**, and shall **not perceive**:

Isaiah the Watchman!

[17] **For verily I say unto you, That many prophets and righteous men have desired to see those things which ye see, and have not seen them; and to hear those things which ye hear, and have not heard them.**

The Passover of Venus had not occurred for generations as they made pilgrimage to visit the kingdom of heaven.

[37] He answered and said unto them, He that **soweth** the **good seed** is the **Son of man**;

The Sun/Son (Venus) of man is responsible for bumper crops.

[38] **The field is the world; the good seed are the children of the kingdom; but the tares are the children of the wicked one;**

The Earth is the battle field as the children of the kingdom (Israel) are laid seige by the tares

*(meteorites) who are the children of the wicked one
Venus (Satan, Beelzebub, Lucifer, etc.). Tares: weeds*

³⁹ The enemy that sowed them is the devil; the harvest is the end of the world; and the reapers are the angels.

*Devil: Venus
Harvest: Destructive Passover
Angel Reapers: Planets/Zodiacs*

⁴⁰ As therefore **the** tares are **gathered** and **burned** in the **fire**; so shall it be in the **end of this world.**

⁴¹ The **Son of man** shall send forth his **angels**, and they shall **gather** out of his kingdom all things that **offend**, and them which do **iniquity**;

The meteorites (tares) from Venus (Sun/Son of man) will bring an end to THIS world. The THIS world that Jesus speaks of is during his time, and not ours!

⁴² And shall **cast** them into a **furnace** of **fire**: there shall be **wailing** and **gnashing** of **teeth**.

⁴³ Then shall the righteous **shine** forth as the **sun** in the **kingdom** of their **Father**. Who hath **ears** to **hear**, let him **hear**.

The furnace in the sky will shine as Son of the Sun in the kingdom of heaven.

*Furnace Victims: Abraham in the Apocalypse of Abraham - Daniel (Belteshazzar), Ashack, Meshack,

and Abednego - Jesus Christ in the Arabic Gospel of the Infancy of the Savior

⁵³ And it came to **pass**, that when **Jesus** had finished these **parables**, he departed thence.

Orbital path change.

⁵⁴ And when he was come into his **own country**, he taught them in their **synagogue**, insomuch that they were **astonished**, and said, Whence hath this man this **wisdom**, and these mighty works?

Chapter 14

⁶ But when **Herod's** birthday was kept, the **daughter** of Herodias **danced** before them, and **pleased** Herod.

Is it not weird to have your stepdaughter do a strip tease for her stepfather?

⁸ And **she**, being before **instructed** of her mother, said, **Give** me here **John Baptist's head** in a **charger**.

This is another example of astrotheology as the head of the lion is held aloft by Orion in the heavens. Nimrod is anthropomorphized as Orion the mighty hunter, and John's head is the star grouping that comprises the lion in the binding.

¹³ When Jesus **heard** of it, he **departed** thence **by ship** into a **desert** place apart: and when the **people**

had **heard** thereof, they **followed** him on **foot** out of the **cities**.

Orbital path change as one does not go by ship to a desert. The people constantly wandered after the Wheel in the Sky (Venus) worshipping (pilgrimage) it as the Kingdom of Heaven!

[15] And when it was **evening**, his disciples came to him, saying, This is a **desert** place, and the **time** is now **past**; send the multitude away, that they may go into the **villages**, and buy themselves **victuals**.

We have another evening event over the desert, and the usual feeding time from the Passover has passed but, still bears food for the masses.

[19] And he **commanded** the multitude to **sit down** on the grass, and took the **five loaves**, and the **two fishes**, and **looking up** to **heaven**, he **blessed**, and **brake**, and gave the **loaves** to his **disciples**, and the **disciples** to the multitude.

Loaves: manna
fishes: Pisces constellation

*We all have seen mussive fish and bird kills during our lives. Would you eat the dead animals? These peoples did, and the people of the Exodus ate irradiated quails and died!

[20] And they did all **eat**, and were **filled**: and they took up of the **fragments** that remained **twelve baskets** full.

12 baskets, 12 disciples, 12 tribes, 12 zodiac signs!

²² And **straightway** Jesus **constrained** his **disciples** to get into a **ship**, and to go before him unto the **other side**, while he sent the multitudes away.

Orbital path change.

²³ And when he had **sent** the multitudes **away**, he went **up** into a **mountain** apart to **pray**: and when the **evening** was come, he was there **alone**.

An Evening Star (Venus) event in a high place.

²⁴ But the **ship** was now **in** the **midst** of the **sea**, **tossed** with **waves**: for the **wind** was **contrary**.

The epic storm force winds blow at the onset of the Passover event.

²⁵ And in the **fourth watch** of the **night Jesus** went **unto** them, **walking** on the **sea**.

We have the rise of the Morning Star Venus, as the Fourth Watch in Jewish time division was 3a.m. to Sun/Son Rise! The stellar jet (privy member, penis, obelisk, etc.) appears to touch the water's surface.

²⁶ And when the disciples **saw** him **walking** on the **sea**, they were **troubled**, saying, It is a **spirit**; and they **cried** out for **fear**.

[27] But **straightway** Jesus **spake** unto them, saying, Be of **good cheer**; it is I; be not **afraid**.

The disciples were afraid that they would killed by the ejection disk, and are being calmed as it is understood by Jesus to be out of range.

[31] And **immediately** Jesus **stretched** forth his **hand**, and caught him, and said unto him, O thou of **little faith**, wherefore **didst** thou **doubt**?

Always Peter.

[32] And when **they** were **come** into the **ship**, the **wind ceased.**

[33] **Then** they that were in the **ship** came and **worshipped** him, saying, Of a truth thou art the **Son of God**.

Venus is recognized as the true Sun/Son of God. This often misinterpreted belief doesn't mesh with our modern sensibilities, as theology, astronomy, and astrology have been synthesized.

[34] And when they were **gone over**, they came into the land of **Gennesaret**.

Orbital path change to Gennesaret.

Chapter 15

²¹ Then Jesus went thence, and departed into the coasts of Tyre and Sidon.

The orbital path heads towards the coasts of Tyre and Sidon in Lebanon.

²³ But he answered her not a word. And his disciples came and besought him, saying, Send her away; for she crieth after us.

²⁴ But he answered and said, I am not sent but unto the lost sheep of the house of Israel.

This reply by Jesus has been a matter of speculation since it was first canonized. The figure known as Jesus Christ was sent (Watcher/Watchman) to teach the populace about the impending disaster from the skies, how to recognize the signs in the skies (kingdom, throne, altar, etc.), cross-contamination prevention, and how to treat the medical issues that will be associated with the Passover event. This warning and resulting ministry (ministering to the heavenly zodiac creatures) were exclusively for the Lost Sheep of Israel that would be destroyed if they did not heed the predictions of Jesus.

²⁵ Then came she and worshipped him, saying, Lord, help me.

²⁶ But he answered and said, It is not meet to take the children's bread, and to cast it to dogs.

This statement of prejudice (Jew versus Gentile) has long been a point of heated debated. The feature in the sky was claimed as the God of the Children of Israel, since the regular parabolic/parabola/parable trajectory brought it into a circumpolar orbit over the land of Israel, thereby making them God's Chosen People, and other nations had to gather (pilgramages) to pay (tributes) homage to the LORD of Hosts (planets, moons, stars, zodiacs).

[27] And she said, Truth, Lord: yet the dogs eat of the crumbs which fall from their masters' table.

The woman turns the understanding around on Jesus!

[29] And Jesus departed from thence, and came nigh unto the sea of Galilee; and went up into a mountain, and sat down there.

Orbital path change over the Sea of Galilee to a high mountain and takes a circumpolar orbit.

[32] Then Jesus called his disciples unto him, and said, I have compassion on the multitude, because they continue with me now three days, and have nothing to eat: and I will not send them away fasting, lest they faint in the way.

We have a clear 3-day orbital event as Venus deposits manna to feed the masses!

[37] And they did all eat, and were filled: and they took up of the broken meat that was left seven baskets full.

The meat (fish) were blown to pieces from the Passover event crossing the water, thereby causing the fish to be explled from the sea, and the loaves are the result of manna (mannAh)!

³⁹ And he sent away the multitude, and took ship, and came into the coasts of Magdala.

Orbital path change over water (ship of heaven) to the coast of Magdala. Magdala in Hebrew means tower or greatness, as it is named after the Wheel in the Sky with Privy Member. The Watchtowers (observatory) were constructed to maintain observation of the heavens. Magdalene means one from Magdala!

Chapter 16

16 The Pharisees also with the Sadducees came, and tempting desired him that he would shew them a sign from heaven.

The Pharisees and Sadducees want to compare notes to see if his observations are accurate.

² He answered and said unto them, When it is evening, ye say, It will be fair weather: for the sky is red.

When the Evening Star (Venus) arrives the sky will be calm, but the skies will bear the crimson (red) flame ahead of the arrival as it burns away the atmosphere of the Earth.

*"Red sky at night Is the sailor's delight;
Red sky in the morning Is a sailor's sure warning."*

³ And in the **morning**, It will be foul weather to day: for the **sky is red and lowering**. O ye **hypocrites**, ye can **discern** the **face of the sky**; but can ye **not** discern the **signs of the times?**

The Morning Star (Venus) rises bringing rain and massive storms as the layers of the atmosphere are peeled away layer by layer. The Pharisees and Sadducees are able to understand the zodiacal bindings, but are not able to discern the impending Passover event from Venus.

⁴ A **wicked** and **adulterous** generation **seeketh** after a **sign**; and there shall **no sign** be **given** unto it, but the sign of the **prophet Jonas**. And he left them, and **departed**.

Jonah's celestial sign is echoed, as the Passover event described by Jonah lasts 48 days and the Passover Tradition is 49!

⁵ And when his **disciples** were come to the **other side**, they had **forgotten** to take **bread**.

No manna (bread/loaves) is deposited with this event.

¹³ When **Jesus** came into the **coasts** of **Caesarea Philippi**, he asked his disciples, saying, **Whom do men** say that **I** the **Son of man** am?

The only way the disciples and Jesus would be able to get to Caesarea Philippi in the ship would be by leaving it! The Sea of Galilee is fed by the Jordan River which is fed by streams from Caesarea Philippi. This temple area is the locale associated with Jesus Christ being given the appellation of the Son/Sun of the living God. This was originally a temple to the Greek God Pan, and Christ shares similar listed attributes with Pan. The declaration of Jesus being validated as a deity at this place speaks to a link of shared theology with so-called pagan beliefs. Caesarea Philippi is also known as Baal-gad (Sun God), Banias (Christian name), Baniyas, Banyas(cult center of Pan), Barias, Belinas, Caesarea Neronias, Caesarea of Philip, Caesarea Paneas (Pan), Caesarea Panias (Pan), Caesareia Sebaste, Keisarion, Kisrin, Medinat Dan, Mivzar Dan, Neronias, Pamias, Paneas (Pan), Paneias (Pan), Paneion (Pan), Panias (Pan), Panium (Pan)!

Temple names: Grotto of the God Pan, Temple of Augustus, Court of Pan & the Nymphs, Temple of Zeus, Court of Nemesis, Temple of the Sacred Goats, and the Temple of Pan & the Dancing Goats - not exactly a Christian or a Jewish site!

PAN-DEMON-ium, PAN-DEMic, PAN-sperm-ia

[18] And I say also unto thee, That thou art Peter, and upon this rock I will build my church; and the gates of hell shall not prevail against it.

Peter is given the proverbial key to the kingdom of heaven. This is wholly astrotheology, as this is a key to a star grouping within the Orion, Sirius, and the Taurus constellations.

[19] And I will give unto thee the keys of the kingdom of heaven: and whatsoever thou shalt bind on earth shall be bound in heaven: and whatsoever thou shalt loose on earth shall be loosed in heaven.

Stars are BOUND (BIND) into the zodiac figures that we are all familiar with.

[21] From that time forth began Jesus to shew unto his disciples, how that he must go unto Jerusalem, and suffer many things of the elders and chief priests and scribes, **and be killed, and be raised again the third day.**

We have the familiar 3-days of Venus during its synod cycle, as 3-days in hell (below the ecliptic).

[23] But he turned, and said unto Peter, Get thee behind me, **Satan**: thou art an offence unto me: for thou savourest not the things that be of God, but those that be of men.

Peter does not believe the event will kill him, and is likened to Satan (Venus) being offensive due to the evil that it did on the Earth.

[24] Then said Jesus unto his disciples, If any man will come after me, let him deny himself, and take up his cross, and follow me.

Jesus had yet to be sentenced to his fate. The cross is a symbol that was used by many cultures prior to the Christian error, like the Babylonian's with the Cross of Tammuz (Sun God - "They weep for Tammuz!").

[27] For the **Son of man** shall **come** in the **glory** of his **Father** with his **angels**; and then he shall **reward** every man according to his **works**.

Venus (Sun of Man) will arrive with meteorites (Angels).

[28] Verily I say unto you, There be some **standing here,** which shall not **taste of death**, till they see the **Son of man** coming in his **kingdom**.

When Venus (Sun of man) comes some of the people there will taste death, and others before then.

Chapter 17

17 And after **six days** Jesus taketh Peter, James, and John his brother, and bringeth them up into an **high mountain apart,**

[2] And was **transfigured** before them: and his face did **shine as the sun, and his raiment was white as the light.**

Venus takes a circumpolar orbit above the mountain, and changes (transfigured) before their eyes. The light cast by Venus during this epoch of time made her shine white like the Sun of man, and this is one of the reasons that Venus was aptly named Lucifer the bringer of light.

Transfigure: to change the outward form or appearance, to change, to glorify or exalt.

Related words: mutate, transmogrify, reform, revise, disfigure, distort, redo, recast, alter

Synonyms: alchemize, make over, metamorphose, convert, transform, transmute, transpose, transubstantiate

Heleph: changing, Passing Over!

[3] And, behold, there ***appeared*** unto them **Moses** and **Elias** talking with him.

[4] Then answered **Peter**, and said unto **Jesus**, Lord, it is **good** for us to be here: if thou **wilt**, let us make here **three tabernacles**; one for thee, and one for **Moses**, and one for **Elias**.

Jesus, Moses, and Elias/Elijah, form the common use of the Trinity of deity.

[5] While he yet **spake**, behold, a **bright cloud overshadowed them**: and behold a **voice out** of the **cloud**, which said, This is my beloved **Son**, in whom I am **well pleased**; hear ye **him**.

Venus continues to transfigure and brighten as the beloved Sun.

*A few cloud names: Abdeel (a **vapor**, a **cloud** of **God**), Abdon/Abaddon/Apollyon (servant, **cloud** of **judgment**!), Admatha (a **cloud** of **death**, a mortal **vapor**), Adoni-bezek (**Lord** of **Lightning**: ex - **Zeus**), Aenon (a **cloud**, **fountain**, his **eye**!), Anani (a **cloud**, prophecy, **divination**), Ananias/Ananiah (the **cloud** of the **Lord**!), Edrei (a very great **mass** or **cloud**!), Enan (**cloud**), Enon (**cloud**, **mass** of **darkness**, fountain, **eye**!), Hazarenan (Imprisoned **cloud**), Ophel (Small white **cloud**)*

I hope that everyone is with me at this point? It is obviously clear what has transpired in antiquity that still echoes misunderstandings into the present!

[6] And when the **disciples** heard it, they **fell** on their **face**, and were **sore afraid**.

[7] And **Jesus** came and **touched them,** and said, **Arise**, and be **not** afraid.

[8] And **when** they had **lifted** up their **eyes**, they **saw** no man, save **Jesus** only.

[9] And as they **came** down from the **mountain**, Jesus **charged** them, saying, **Tell** the **vision** to no man, until the **Son** of **man** be **risen again** from the **dead**.

The disciples cringe in fear as they worry about destruction from the cloud (Venus) as the Sun of man will be resurrected (risen, reborn) from below the ecliptic (Hell).

¹⁰ And his **disciples** asked him, saying, Why then **say** the **scribes** that **Elias** must **first** come?

¹¹ And **Jesus** answered and said **unto** them, Elias **truly** shall first **come**, and **restore** all things.

¹² **But I say unto you, That Elias is come already, and they knew him not, but have done unto him whatsoever they listed. Likewise shall also the Son of man suffer of them.**

¹³ Then the **disciples** understood that he **spake** unto **them** of **John the Baptist**.

Elias/Elijah the Watchman and John the Baptist have been commonly seen to two separate people and times.

¹⁵ **Lord**, have **mercy** on my **son**: for he is **lunatick**, and sore **vexed**: for **ofttimes** he **falleth** into the **fire**, and **oft** into the **water**.

I believe this to be more about the apects of Venus on her Passovers, as she would sometimes deliver fire, and other times water (Sea of Galilee?).

Lunatic: Moon worshipper Luna: Moon

²² And while they abode in **Galilee**, Jesus said unto them, The **Son of man** shall be **betrayed** into the hands of men:

²³ And they shall kill him, and **the third day** he shall be raised again. And they were exceeding sorry.

Venus 3-day synod cycle.

²⁴ And **when** they were come to **Capernaum**, they that **received** tribute **money** came to **Peter**, and said, Doth not your **master** pay **tribute**?

The orbital path goes from Galilee to Capernaum.

Chapter 18

¹⁰ Take **heed** that ye **despise** not one of these **little ones**; for I say **unto** you, That in **heaven** their **angels** do always **behold** the **face** of my **Father** which is in **heaven**.

This verse drips of astrotheology, as we are told not to hate (despise) the zodiac stars (little angels) as they always orbit (look) around (face) the Sun (Father) in the zodiac bindings.

¹¹ For the **Son of man** is **come** to **save** that **which** was **lost**.

Sun of man = Venus

¹⁴ Even **so** it is **not** the **will** of your **Father** which is in **heaven**, that **one** of these **little** ones should **perish**.

We are led to understand that the zodiac bindings will not perish as the people fear.

18 Verily I say **unto** you, **Whatsoever** ye shall **bind** on **earth** shall be **bound** in **heaven**: and whatsoever ye shall **loose** on **earth** shall be **loosed** in **heaven**.

Bind = zodiac bindings (stabilizing the stars)
Heaven/Earth = As Above/So Below

22 Jesus saith unto him, I say not unto thee, Until seven times: but, Until seventy times seven.

Another Passover event in 490 years from that time? Seventy were the number of children/constellations of the Canaanite God El (Saturn), and seven is Venus/Inna's sacred number.

Seventy: constellations
Seven: known planets at that time

23 Therefore is the kingdom of heaven likened unto a certain king, which would take account of his servants.

WOW! We are clearly being shown that in parables the celestial heavens (kingdom of heaven) are being anthropomorphized to facilitate a far simpler conveyance of understanding. The certain king would be the Sun, and the 70 constellations are the servants!

Chapter 19

1 And it came to pass, that when Jesus had finished these sayings, he departed from **Galilee**, and came into the coasts of **Judaea** beyond **Jordan**;

Orbital path change from Capernaum to Galilee and then on to Judea.

18 He saith **unto** him, Which? **Jesus** said, **Thou** shalt do **no murder, Thou** shalt **not commit adultery, Thou** shalt **not steal,** Thou shalt **not bear false witness,**

19 Honour thy **father** and thy **mother**: and, **Thou** shalt **love** thy **neighbour** as thyself.

All of these are excellent standards to live by.

28 And **Jesus** said unto them, **Verily** I say unto **you,** That ye **which** have **followed me,** in the **regeneration** when the **Son of man** shall **sit** in the **throne of his glory,** ye also shall **sit** upon **twelve thrones, judging** the **twelve tribes** of Israel.

Jesus relates that the 12 zodiac constellations that have followed the Sun of man in the heavens, will judge the 12 zodiac constellations of the Old Testaments Children of Israel. When the 12 zodiac constellations of the New Testament regenerate the 12 zodiac constellations the healing will be complete.

Chapter 20

1 For the **kingdom of heaven** is like **unto** a **man** that is an **householder,** which **went** out **early** in the **morning** to hire **labourers** into his **vineyard**.

Venus as the Morning Star appears to help the vineyards growth with Passovers.

17 And **Jesus** going up to **Jerusalem** took the **twelve disciples** apart in the **way**, and said unto them,

The orbital event moves to Jerusalem.

18 **Behold**, we go up to **Jerusalem**; and the **Son of man** shall be **betrayed** unto the **chief priests** and unto the **scribes**, and they shall **condemn** him to **death**,

The Sun/Son of man will be betrayed (different astronomical observations) by the chief priests and scribes. We see that Jesus is opposed by the science-literary arm of the Pharisees and Sadducees, as they do not believe in the strength or significance of the Passover to come, and do not feel that it is the Sun/Son of man.

19 And shall **deliver** him to the **Gentiles** to **mock**, and to **scourge**, and to **crucify** him: and the **third day** he shall **rise** again.

Always the Jews versus Gentiles theme. We have the common 3-day resurrection cycle of Venus below the visible ecliptic plane.

29 And as they **departed** from **Jericho**, a **great** multitude **followed** him.

Orbital path from Jerusalem to Jericho.

33 They **say** unto him, **Lord**, that our **eyes** may be **opened**.

34 So Jesus had compassion on them, and touched their eyes: and immediately their eyes received sight, and they followed him.

Healing of the blind was the most common cure that Jesus is associated with, as the comet/proto-planet Venus may have been the cause and the cure of the eye issues noted. All of the blindness Miracles occur in accompaniment with water or soil near the Sea of Galilee (DMP) applied directly to the sufferers eyes. People flock to the Dead Sea (DMP), Sea of Galilee (DMP), and to Lourdes in France (DMP?) to partake of the miraculous waters, salt, and soil. Staring directly at a stellar jet/nuclear reaction will cause blindness. It is noted that the Children of the Exodus walked with their eyes lowered when God Passed Over them on their 40 year sojourn. This would make excellent sense, as they would have learned prior that this was required in order to avoid death or blindness.

Chapter 21

1 And when they **drew** nigh unto **Jerusalem**, and were come to **Bethphage**, unto the **mount of Olives**, then sent Jesus **two disciples**,

Orbital path change from Jericho to Bethphage with two zodiacs signs following.

12 **And Jesus went into the temple of God, and cast out all them that sold and bought in the temple, and overthrew the tables of the moneychangers, and the seats of them that sold doves,**

The Passover destroys (plasma discharge?) the synagogue/temple of God. This often spoken of passage has shown Jesus storming the building and tossing tables. The moneychangers, Romans, and Jewish leadership would have had him strung up immediately, as they would have taken the revenue offenses as treason! Doves (sacrifices?)?

[14] And the **blind** and the **lame** came to him in the **temple**; and he **healed** them.

The Miracles flow after each Passover event!

[17] And he **left** them, and **went** out of the **city** into **Bethany**; and he **lodged** there.

Orbital path change Bethphage to Bethany and take a circumpolar orbit.

[18] Now in the morning as he returned into the city, he hungered.

We have a Morning Star event.

[21] Jesus answered and said **unto** them, **Verily** I say unto you, If **ye** have **faith**, and **doubt** not, ye shall not **only** do this which is **done** to the **fig tree**, but also if ye shall say unto this **mountain**, Be thou **removed**, and **be** thou **cast** into the **sea**; it **shall** be **done**.

Jesus is trying to teach the disciples that if they have faith in the scientific findings, then the mountain (Venus) will fall into the sea based on the trajectory.

Chapter 22

1 And **Jesus** answered and **spake** unto them **again** by **parables**, and said,

Parables aka parabola (parabolic).

[23] The **same day** came to him the **Sadducees**, which say that **there is no resurrection**, and asked him,

The Sadducees do not believe that Venus will rise from below the ecliptic in hell to be a threat again.

[32] I am the **God of Abraham**, and the **God of Isaac**, and the **God of Jacob**? God is **not** the **God of the dead**, but of the **living**.

The living God of Abraham (Abram/Brama) & Isaac was the destructive God of antiquity as Venus was more active and potent during that period, with her parabolic trajectory brought her closer to the Earth. The living God of the New Testament was one of Salvation and Miracles, as Venus had begun to lose her features from Passovers in antiquity.

Chapter 23

[7] And **greetings** in the **markets**, and to be **called** of men, **Rabbi, Rabbi**.

Rabbi = Hammurabi? The Law Code of Hammu-RABBI is the first known law codes from Sumeria that still survive today.

Chapter 24

[2] And **Jesus** said unto **them**, See ye not all **these things**? verily I **say** unto you, There shall **not** be **left** here **one** stone upon **another**, that **shall** not **be** thrown **down**.

The Passover will bring structures down in the city from earthquakes.

[3] And as he sat upon the **mount of Olives**, the disciples **came** unto him **privately**, saying, Tell us, **when** shall these things **be**? and what **shall** be the **sign** of thy **coming**, and of the **end of the world**?

The disciples want to know what features (sign) will appear in the sky when the Passover event occurs. The end of the world did not come for all of them, and this will not happen for quite some time.

[7] For **nation** shall rise **against** nation, and **kingdom** against **kingdom**: and there shall be **famines**, and **pestilences**, and **earthquakes**, in **divers** places.

These Passover events brought upheaval with drought and desertification, as kingdoms and land ownership changed hands constantly. The earthquakes were a result of gravitational pull from circumpolar orbits.

[14] And this **gospel** of the **kingdom** shall be **preached** in all the **world** for a **witness** unto all **nations**; and then **shall** the **end** come.

Misinterpreted verses like this one, have had humanity awaiting the end since the beginning.

¹⁶ Then let them which be in Judaea flee into the mountains:

When the comet Venus is not in the regular orbital path, then those in Judea must head to the mountains for safety. John of Patmos relates that the mountains and caves became their tombs!

¹⁷ Let him which is on the housetop not come down to take any thing out of his house:

¹⁸ Neither let him which is in the field return back to take his clothes.

Jesus has to issue evacuation instructions as F.E.M.A. would. Isaiah warned them in the Old Testament to stop following the Lord and to get off of the path (DMP)!

¹⁹ And woe unto them that are with child, and to them that give suck in those days!

This has been noted by many other authors, as the state of the land caused famine bringing breast milk production to an all time low.

²⁰ But pray ye that your flight be not in the winter, neither on the sabbath day:

Naked in the snow on a Sunday (Sun Day - Christian)/Saturday (Saturn's Day - Judaism) is not a good way to evacuate.

[22] And **except** those **days** should be **shortened**, there **should** no **flesh** be **saved**: but for the **elect's** sake those **days** shall be **shortened**.

Jesus is telling us that due to the orbital trajectory, the earth will slow down from the Passover. The Sundial in Ahaz went backwards 10°'s in Isaiah's days, noted as an extended period of sunlight.

[26] Wherefore if they shall say **unto** you, **Behold**, he is in the **desert**; go **not forth**: behold, **he is** in the **secret chambers**; believe it not.

The understanding is that when Venus goes below the ecliptic in the desert, do not believe that the event has Passed Over. The 3-day synod cycle in the secret chambers will allow it to be reborn.

[27] For as the lightning cometh out of the east, and shineth even unto the west; so shall also the coming of the Son of man be.

The lightning of Venus (Adonai-bezek) will arrive from the East (Venus is reborn as the Morning Star) and shines brighter than the Sun in the sky as the Sun/Son of man.

[29] Immediately after the **tribulation** of those **days** shall the **sun be darkened**, and the **moon shall not**

give her light, and the **stars** shall **fall from heaven**, and the **powers** of the **heavens** shall be **shaken**:

A transit of Venus across the face of the Sun typically will occur (rare event) during the month of December (winter!).Tthe Moon is then blocked with meteors falling to Earth from the Pass Over as the heavens are damaged from the event.

[30] And **then** shall **appear** the **sign** of the **Son of man in heaven**: and then shall **all** the **tribes** of the **earth mourn**, and they shall **see** the **Son of man** coming in the **clouds** of **heaven** with **power** and **great glory**.

As the approach of Venus begins the sign is seen that the Sun of man has come in the clouds! Remember all of the names of cities associated with clouds!

[31] And he shall **send** his **angels** with a **great sound** of a **trumpet**, and they shall **gather** together his **elect** from the **four winds**, from **one end** of **heaven** to the other.

The fallen angels (meteorites) arrive as the skies reverberate from the Passover of Venus.

[32] Now **learn** a **parable** of the **fig tree**; When his **branch** is yet **tender**, and **putteth** forth leaves, ye know that **summer** is **nigh**:

[33] So **likewise** ye, when ye **shall** see all **these things**, know **that** it is near, **even** at the **doors**.

Jesus represents a harvest god.
U.S. Spring - @ March 21 to June 21
Summer - @ June 22 to August 21
Passover tradition: April 15

34 Verily I say unto you, **This generation** shall **not pass**, till **all** these **things** be **fulfilled**.

We have had 80 generations since the time of Christ!

35 Heaven and **earth** shall **pass away**, but my **words** shall not **pass away**.

The words of his teachings and misunderstandings will resonate.

36 But of that **day** and **hour** knoweth **no man**, no, not the **angels** of **heaven**, but my **Father only**.

The planets and stars do not know when the LORD of hosts will destroy the world.

38 For as in the days that were before the flood they were eating and drinking, marrying and giving in marriage, until the day that Noe entered into the ark,

We have Venus as the destroyer of record for the Great Flood @ 2344 BCE.

40 Then shall **two** be in the **field**; the **one** shall be **taken**, and the **other left.**

One man will live, and one man will die from the Passover.

⁴² Watch therefore: for ye know not what hour your Lord doth come.

With the Morning Star!
Watch=Watcher=Watchman=Watch Tower

⁴³ But know this, that if the goodman of the house had known in what watch the thief would come, he would have watched, and would not have suffered his house to be broken up.

The Goodman is a Watchman from the north and the New Testament. Venus the thief would have probably come in the 4th watch in the dawn hours.

⁵¹ And shall cut him asunder, and appoint him his portion with the hypocrites: there shall be weeping and gnashing of teeth.

The human life span is 72 years and is considered your portion. A celestial portion (1° = 72 yrs x 360 = 25,920 yrs = A full zodiac cycle) is what this verse refers to, as our time will be cut short during the time of Jesus.

Chapter 25

³³ And he shall set the sheep on his right hand, but the goats on the left.

The evil goat would be placed on the left, and the angelic sheep on the right. This ritual sacrifice practice was used by the Children of the Exodus in their Scapegoat practice. The left side is considered evil, and the right side is considered angelic.

Chapter 26

[2] Ye know that after **two days is the feast of the passover**, and the **Son of man** is **betrayed** to be **crucified**.

The Passover and Crucifixion event is celebrated April 15th.

[6] Now when Jesus was in **Bethany**, in the **house of Simon the leper**,

Orbital path change to Bethany.

[15] And said unto them, What will ye give me, and I will deliver him unto you? And they **covenanted** with him for **thirty** pieces of silver.

Covenant: Agreement/pact usually with a God or deity
Cove: restricted waterway
Coven: 13 Witches
Covenant = Witches agreement!
30 pieces of silver=30 days

[18] **And he said, Go into the city to such a man, and say unto him, The Master saith, My time is at hand;**

I will keep the passover at thy house with my disciples.

This statement echoed by Luke and others is wholly astrotheology! The city that they enter is the celestial heavenly city of the zodiac, as the man carrying the water is Aquarius (2012 CE to 4172 CE) who follows the sign of Pisces (148 BCE to 2012 CE!) in the zodiac bindings.

*Luke 22:10 And he said unto them, Behold, when ye are entered into the city, there shall a man meet you, bearing a pitcher of water; follow him into the house where he entereth in.

Aquarius, the man carrying the water follows Jesus as Pisces in the zodiac bindings.

[19] And the **disciples** did as **Jesus** had **appointed** them; and they **made** ready the **passover**.

The event is set to Passover, and is typically celebrated April 14th or the first full moon in the month of Nisan.

[20] **Now when the even was come, he sat down with the twelve.**

[26] And as they were **eating**, Jesus took **bread**, and **blessed** it, and **brake** it, and gave it to the disciples, and said, Take, **eat**; this **is** my **body**.

[27] And he took the **cup**, and gave **thanks**, and gave it to them, saying, **Drink** ye all of it;

28 For this is my **blood** of the **new testament**, which is **shed** for many for the **remission** of **sins**.

29 But I say unto you, I will not **drink henceforth** of this **fruit** of the **vine**, until that day when I **drink** it new with you in my **Father's kingdom**.

These verses is an allegory to the healing of the heavens that will take place after the Passover event. Jesus was attempting to give the disciples a scientific understanding through symbolism. This symbolism has been retaught improperly since it was first done and reminds one of a pagan ritual practice (blood-wine/bread=manna).

30 And **when** they had **sung** an **hymn**, they went **out** into the **mount of Olives.**

Orbital path change to the mount of Olives.

32 But after I am **risen again, I will go before you into Galilee.**

When Venus is resurrected from the 3-day synod cycle the orbital path will take it over Galilee.

34 Jesus said unto him, Verily I say unto thee, That this **night**, before the **cock crow**, thou shalt deny me thrice.

36 Then cometh Jesus with them unto a place called **Gethsemane**, and saith unto the disciples, Sit ye here, while I go and **pray yonder.**

Orbital path change from the Mount of Olives to Gethsemane.

⁴⁰ And he **cometh** unto the disciples, and **findeth** them **asleep**, and saith unto **Peter**, What, could ye **not watch** with me **one hour**?

1st ejection disk (privy member) groundfall. A Watchman asleep on duty!
First Watch: Sundown to 9pm.

⁴¹ **Watch and pray**, that ye **enter** not into **temptation**: the **spirit** indeed is **willing**, but the **flesh** is **weak**.

Pray that the path of Venus the Tempter (Satan) whose spirit is willing to destroy the weak flesh Passes over.

⁴² He went **away** again the **second time**, and prayed, saying, O my **Father**, if this cup may not **pass away** from me, except I **drink** it, thy will be done.

2nd ejection disk (privy member) ground fall.
Second Watch: 9pm to midnight.

⁴⁴ And he **left** them, and went **away** again, and **prayed** the **third time**, saying the same words.

3rd ejection disk (privy member) ground fall.
Third Watch: Midnight to 3am.

⁴⁵ Then **cometh** he to his **disciples**, and saith unto them, **Sleep** on now, and **take** your **rest**: behold, the

hour is at **hand**, and the **Son of man** is **betrayed** into the **hands** of **sinners**.

The event is set to take place.
Fourth Watch: 3 am to Sunrise (cock crow!)

46 Rise, let us be **going**: behold, **he** is at **hand** that **doth** betray me.

53 Thinkest thou **that** I **cannot** now pray to my **Father**, and he shall **presently** give me more than **twelve legions of angels?**

Twelve legions of twelve zodiac bindings!

1 Roman Legion = 5,120 soldiers X 12 = 61,440

54 But how then **shall** the **scriptures** be **fulfilled**, that thus it **must** be?

The scriptures were the orbital paths noted and predicted in the future by Watchman like Isaiah and Jeremiah.

56 But all this was **done**, that the **scriptures of the prophets might be fulfilled**. Then all the **disciples forsook him**, and **fled**.

This speaks to all of the visible stars and planets removed from sight.

61 And said, This fellow said, I am able to destroy the temple of God, and to build it in three days.

The synagogues were destroyed during a 3-day synod cycle.

⁶⁴ Jesus saith unto him, Thou hast said: nevertheless I say unto you, Hereafter shall ye see the Son of man sitting on the right hand of power, and coming in the clouds of heaven.

Jesus declares that his astronomical observations are accurate, and that they will see Venus in the clods of heaven on the right (Good side) hand of God (Sun).

⁷⁴ Then began he to curse and to swear, saying, I know not the man. And immediately the cock crew.

⁷⁵ And Peter remembered the word of Jesus, which said unto him, Before the cock crow, thou shalt deny me thrice. And he went out, and wept bitterly.

The cock crows with the rising of the Sun and the Morning Star (Venus) sitting on the right hand side just as his astronomical observations predicted.

Chapter 27

1 When the morning was come, all the chief priests and elders of the people took counsel against Jesus to put him to death:

² And when they had bound him, they led him away, and delivered him to Pontius Pilate the governor.

This speaks of star binding as well as human binding.

²² Pilate saith unto them, What **shall** I do then with **Jesus** which is **called Christ? They** all **say** unto him, Let **him** be **crucified**.

Pilate refers to Jesus as is called Christ, and he is crucified on the Southern Crux/Cross constellation. Many other deities have suffered the same fate as Jesus well before his time.

³² And as they came out, they found a man of **Cyrene**, **Simon** by name: him they compelled to bear his cross.

Cyrene = Libyan

³³ And when they were come unto a place called **Golgotha**, that is to say, a **place of a skull,**

Golgotha: Heap of a skull, or skulls
Goliath of Goth = Gol/goth/a

Goliath was a celestial figure as the head of the lion that Orion (David) holds aloft in the heavens.

³⁵ And they **crucified** him, and **parted** his garments, **casting lots**; that it might be fulfilled which was **spoken** by the prophet, They **parted** my garments among them, and upon my **vesture** did they **cast lots**.

The casting of lots is known as cleromancy, and uses bones, rocks, or dice, for divination.

38 Then were there two thieves crucified with him, one on the right hand, and another on the left.

The Southern Crux/Cross constellation star on the right hand of the cross is Becrux (Good) and the left side is Delta Cru (Evil) and Epsilon Cru in the middle. The alignment may also refer to the 3-Wise Men in the Belt of Orion.

45 Now from the sixth hour there was darkness over all the land unto the ninth hour.

We have a total eclipse from a Passing planetary body from noon (Sixth hour - noon to 1pm) until 3pm (Ninth hour - 3pm to 4pm) lasting about 3 hours in total.

46 And about the ninth hour Jesus cried with a loud voice, saying, Eli, Eli, lama sabachthani? that is to say, My God, my God, why hast thou forsaken me?

Eli = El = Canaanite God = Saturn

47 Some of them that stood there, when they heard that, said, This man calleth for Elias.

Elias = Elijah = Watchman

51 And, behold, the veil of the temple was rent in twain from the top to the bottom; and the earth did quake, and the rocks rent;

The Passover event occurs as the temple cracks and the massive earthquake causes destruction.

⁵² And the graves were opened; and many bodies of the saints which slept arose,

Land sinks and rises causing graves to give up the dead. Events like this led to the superstitious belief of the dead rising to kill the living, and which is why zombies are so popular today.

⁵³ And came out of the graves after his resurrection, and went into the holy city, and appeared unto many.

We have the completion of the 3-day synod cycle.

⁵⁴ Now when the centurion, and they that were with him, watching Jesus, **saw the earthquake**, and those things that were done, they feared greatly, saying, **Truly this was the Son of God.**

Venus is recognized by a Roman centurion as the Sun of God.

⁵⁵ And many women were there beholding **afar off**, which followed **Jesus from Galilee**, ministering unto him:

The orbital path is followed from Golgotha (Heap of skulls) to Galilee (wheel, cylinder, circuit, district) by the people.

Chapter 28

28 In the **end of the sabbath**, as it began to **dawn** toward the **first day of the week**, came Mary Magdalene and the other Mary to see the sepulchre.

Sabbath: Saturday (Hebrew - Saturn's Day) & Sunday (Christian - Sun Day)
Dawn: Morning Star - Venus

² And, behold, there was a **great earthquake**: for the **angel of the Lord descended from heaven, and came and rolled back the stone from the door, and sat upon it.**

The resurrection of Venus as the Morning Star brings a massive earthquake (not noted in the EQ records)with her appearance.

³ **His countenance was like lightning, and his raiment white as snow:**

Lightning and electricity crackles as the massive white hot planet Passes Over.

⁷ **And go quickly, and tell his disciples that he is risen from the dead; and, behold, he goeth before you into Galilee; there shall ye see him: lo, I have told you.**

We have the orbital path heading to Galilee as the populace follows in adoration and worship.

¹¹ Now **when** they were **going**, behold, **some of the watch** came **into** the **city**, and **shewed** unto the **chief priests** all the **things** that were **done**.

The chief priests that were awakened to the Pass Over are informed by their Watchmen that the observations of Jesus were accurate.

[16] Then the **eleven disciples** went away into **Galilee, into a mountain** where Jesus had appointed them.

One zodiac binding (disciple) is destroyed as Venus takes a circumpolar orbit above a mountain in Galilee.

[20] **Teaching them to observe all things whatsoever I have commanded you: and, lo, I am with you always, even unto the end of the world (Age/Aeon). Amen.**

We close the Book of Matthew with a wholly astrotheological statement! The early Christians were firm believers in the close or end of an age. The age speaks of zodiac ages, as Jesus lets them know that his sign of Pisces will reign until (2012 CE) when the AGE or commonly understood WORLD will come to a end, and will usher in the man carrying the water Aquarius!

Neo = New   Eon = Age   Neo Eon = New Age!

* Reprinted in part from; "Solomon, Isaiah, Jesus Christ, and the Biblical Disaster-Miracle Path"

THE ARABIC GOSPEL OF THE INFANCY OF THE SAVIOUR

3. Wherefore, **after sunset**, the old woman, and Joseph with her, came to the cave, and they both went in. And, behold, it was filled with **lights** more beautiful than the **gleaming** of **lamps** and **candles**, (4) and more splendid than the **light of the sun**. The child, enwrapped in swaddling clothes, was sucking the breast of the Lady Mary His mother, being placed in a **stall**. And when both were wondering at this light, the old woman asks the Lady Mary: Art thou the mother of this Child? And when the Lady Mary gave her assent, she says: Thou art not at all like the **daughters of Eve.** The Lady Mary said: As my son has no equal among children, so his mother has **no equal** among women. The old woman replied: My mistress, I came to get payment; I have been for a long time affected with **palsy**. Our mistress the Lady Mary said to her: **Place thy hands upon the child**. And the old woman did so, and was immediately cured. Then she went forth, saying: Henceforth I will be the **attendant** and **servant** of this child all the days of my life.

Fresh into the world and the miracles are beginning.

The birth of Christ on four calendar systems: Hebrew-1st day of Passover, Sacred Round –Reed, Venus – Creation, Mercury - Creation

Born: December 25th - 355° solar point
Conception: March 15th - 75° solar point

Symbolism: March 21st - 81° solar point marks the onset of Spring and planting

4. Then came **shepherds**; and when they had lighted a fire, and were rejoicing greatly, there appeared to them the **hosts of heaven** praising and celebrating **God Most High**. And while the shepherds were doing the same, the cave was at that time made like a **temple of the upper world**, since both **heavenly** and **earthly** voices glorified and magnified God on account of the birth of the Lord Christ. And when that old Hebrew woman saw the **manifestation** of those **miracles**, she thanked God, saying: I give Thee thanks, O God, the **God of Israel**, because mine eyes have seen the birth of the Saviour of the world.

Shepherds (constellations), hosts of heaven (planets & stars), temple of the upper world (wow!), heavenly (As above) and Earthly (So below)! The astrotheology oozes from this statement as it is clear that we are based in the stars.

6. Then old Simeon saw Him **shining like a pillar of light**, when the Lady Mary, His virgin mother, rejoicing over Him, was carrying Him in her arms. And **angels**, praising Him, stood **round** Him in a **circle**, like life guards standing by a king. Simeon therefore went up in haste to the Lady Mary, and, with hands stretched out before her, said to the Lord Christ: Now, O my Lord, let Thy servant depart in peace, according to Thy word; for mine eyes have seen Thy compassion, which Thou hast prepared for the salvation of all peoples, a light to all nations, and glory to Thy people Israel. **Hanna** also, a **prophetess,**

was present, and came up, giving thanks to God, and calling the Lady Mary blessed. (4)

This is more astrotheology espoused in Biblical context. The shining like a pillar of light (fiery wheel in the sky), angels (An & El), round in a circle (the zodiac constellations are bound in a circular chariot always on guard!).

7. And it came to pass, when the **Lord Jesus** was born at **Bethlehem of Judaea**, in the time of **King Herod**, behold, **magi** came from the **east** to **Jerusalem**, as **Zeraduscht** (5) had **predicted**; and there were with them gifts, gold, and frankincense, and myrrh. And they adored Him, and presented to Him their gifts. Then the **Lady Mary** took one of the swaddling-bands, and, on account of the smallness of her means, gave it to them; and they received it from her with the greatest marks of honour. And in the **same hour** there **appeared** to them an **angel in the form of that star** which had before **guided** them on their journey; and they went away, following the **guidance of its light**, until they arrived in their own country. (6)

Magi, east, predicted, hour of appearance, star, and guidance of its light!

8. And their kings and chief men came together to them, asking what they had seen or done, how they had gone and come back, what they had brought with them. And they showed them that swathing-cloth which the Lady Mary had given them. Wherefore they celebrated a feast, and, according to

their custom, lighted a fire and worshipped it, and threw that swathing-cloth into it; and the fire laid hold of it, and enveloped it. And when the fire had gone out, they took out the **swathing-cloth exactly as it had been before**, just as if the fire had not touched it. Wherefore they began to **kiss it,** and to put it on their heads and their eyes, saying: This verily is the **truth without doubt**. Assuredly it is a great thing that the fire was not able to burn or destroy it. Then they took it, and with the greatest honour laid it up among their treasures.

The swathing-cloth comes out unscathed just as Abram did from the furnace.

9. And when Herod saw that the **magi** had left him, and not come back to him, he summoned the **priests** and the **wise men**, and said to them: Show me where Christ is to be born. And when they answered, In Bethlehem of Judaea, he began to think of putting the Lord Jesus Christ to death. Then appeared an angel of the Lord to Joseph in his sleep, and said: Rise, take the boy and His mother, and go away into **Egypt**. (7) He **rose**, therefore, **towards cockcrow**, and set out.

Magi (magicians), priests, and wise men are separated into three distinct groups! The East (rising sun) is towards where the cockcrow! If they were heading to Egypt, they would be going towards the west and not towards the cockcrow as the approaching sunrise would be at their backs.

The trip to Egypt of Christ in relation to five calendar systems: Hebrew – Mourning, Sacred Round – Dragon, Venus – Birth, Mercury – Resurrection!

10. While he is reflecting how be is to set about his journey, morning came upon him after he had gone a very little way. And now he was approaching a great city, in which there was an idol, to which the other idols and gods of the Egyptians offered gifts and vows. And there stood before this **idol** a priest **ministering** to him, who, as often as **Satan** spoke from that idol, reported it to the inhabitants of Egypt and its territories. This priest had a son, **three years old**, beset by **several demons**; and he made many speeches and utterances; and when the demons seized him, he tore his clothes, and remained naked, and threw stones at the people. And there was a hospital in that city dedicated to that idol. And when Joseph and the Lady Mary had come to the city, and had turned aside into that hospital, the **citizens** were very much **afraid**; and all the chief men and the priests of the idols came together to that idol, and said to it: What **agitation** and **commotion** is this that has **arisen** in our **land**? The **idol answered them:** A **God** has come here in secret, who is God **indeed**; nor is any god besides Him worthy of divine worship, because He is truly the Son of God. And when this land became **aware** of His **presence**, It **trembled** at **His arrival**, and was **moved** and **shaken**; and we are **exceedingly afraid** from the greatness of His power. And in the **same hour** that **idol fell down**, and at its fall all, inhabitants of **Egyp**t and **others, ran together**.

We would have someone in a straightjacket if they said that they audibly heard an inanimate object talk in prophecy, yet we do not hold these people of antiquity under the same microscope! Demons, citizens afraid, agitation and commotion, ARISEN in our land, presence, trembled, arrival, moved, shaken, exceedingly afraid, same hour, idol fell down, Egypt, others ran together. This sounds like the all too familiar Passover of a heavenly body arising over the land of Egypt causing the Earth to move and shake, bringing down the idols and other structures in the region.

11. And the son of the priest, his usual disease having come upon him, entered the hospital, and there came upon Joseph and the Lady Mary, from whom all others had fled. The Lady Mary had washed the cloths of the Lord Christ, and had spread them over some wood. That **demoniac boy**, therefore, came and took one of the cloths, and put it on his **head**. Then the demons, **fleeing** in the shape of **ravens** and **serpents**, began to go forth out of his **mouth**. The boy, being immediately healed at the command of the Lord Christ, began to praise God, and then to give thanks to the Lord who had healed him. And when his father saw him restored to health, My son, said he, what has happened to thee? and by what means hast thou been healed? The son answered: When the demons had thrown me on the ground, I went into the hospital, and there I found an **august** woman with a boy, whose newly-washed cloths she had thrown upon some wood: one of these I took up and put upon my head, and the demons left me and fled. At this the father rejoiced greatly, and said: My son, it is possible that this boy

is the Son of the living God who created the **heavens and the earth**: for when he came over to us, the idol was broken, and all the gods fell, and perished by the power of his magnificence.

Ravens: Corvus constellation
Serpents: Serpens constellation
Draco constellation
Water Serpent: Hydra constellation

12. Here was fulfilled the prophecy which says, Out of Egypt have I called my son. (1) Joseph indeed, and Mary, when they heard that that idol had fallen down and perished, trembled, and were afraid. Then they said: When we were in the land of Israel, Herod thought to put Jesus to death, and on that account **slew** all the **children** of **Bethlehem** and its confines; and there is no doubt that the Egyptians, as soon as they have heard that this idol has been broken, will burn us with fire. (2)

Herod slaughters all the children as Nimrod only slew one trying to kill Abram.

The trip from Egypt corresponds on four calendar systems: Hebrew - Consecration, Sacred Round – Dragon, Venus – DEATH, Merucry - Creation

13. Going out thence, they came to a place where there were robbers who had plundered several men of their baggage and clothes, and had bound them. Then the robbers heard a great noise, like the noise of a magnificent king going out of his city with his army, and his **chariots** and his drums; and at this the

robbers were terrified, and left all their plunder. And their captives rose up, loosed each other's bonds, recovered their baggage, and went away. And when they saw Joseph and Mary coming up to the place, they said to them: Where is that **king**, at the hearing of the magnificent sound of whose approach the robbers have left us, so that we have escaped safe? Joseph answered them: **He will come behind us**.

A two-year old following on his own? They speak of a heavenly body (Sun) behind them.

15. On the **day after**, being supplied by them with provision for their journey, they went away, and on the **evening** of that **day** arrived at another town, in which they were celebrating a marriage; but, by the arts of accursed Satan and the work of enchanters, the bride had become dumb, and could not speak a word. And after the Lady Mary entered the town, carrying her **son the Lord Christ**, that **dumb bride** saw her, and stretched out her hands towards the Lord Christ, and drew Him to her, and took Him into her arms, and held Him close and kissed Him, and leaned over Him, moving His body back and forwards. Immediately the knot of her tongue was **loosened**, and her ears were opened; and she gave thanks and praise to God, because He had restored her to health. And that night the inhabitants of that town exulted with joy, and thought that God and His angels had come down to them.

Lord Christ: Infant Prince constellation
Dumb Bride: Andromeda constellation

16. There they **remained three days**, being held in great honour, and living splendidly. Thereafter, being supplied by them with provision for their journey, they went away and came to another city, in which, because it was very populous, they thought of passing the night. And there was in that city an excellent woman: and once, when she had gone to the **river** to bathe, lo, accursed **Satan**, in the form of a **serpent**, had leapt upon her, and **twisted** himself round her **belly**; and as often as **night came on**, he **tyrannically tormented** her. This woman, seeing the mistress the **Lady Mary**, and the child, the Lord Christ, in her bosom, was struck with a longing for Him, and said to the mistress the Lady Mary: O mistress, give me this child, that I may carry him, and kiss him. She therefore gave Him to the woman; and when He was brought to her, Satan let her go, and fled and left her, nor did the woman ever see him after that day. Wherefore all who were present praised God Most High, and that woman bestowed on them liberal gifts

The common 3 day descent below the ecliptic plane as this speaks to an orbital path passing the night in another city/house. Satan speaks of Venus, and the serpent is the Serpens constellation that wraps 1/3 around the heavens and sits above the Mother and Child (Infant Prince Constellation) in the zodiac binding. The Andromeda constellation (dumb bride) takes hold of the Infant Prince and this causes Satan (Venus) to release her and flee its' orbit.

51. And a **philosopher** who was there present, a **skilful astronomer**, asked the Lord Jesus whether He had studied astronomy. And the Lord Jesus

answered him, and explained the **number of the spheres**, and of the **heavenly bodies, their natures and operations; their opposition; their aspect, triangular, square, and sextile; their course, direct and retrograde; the twenty-fourths,(2) and sixtieths of twenty-fourths**; and other things beyond the reach of reason.

We see the whole of the understanding, as Jesus is a philosopher, astronomer, and mathematician. This would be expected of one that attended the Egyptian Mystery School, as they focused on math, science, and speaking. The Arabic Gospel dates to the fifth or sixth century CE, and shows that the whole of the understanding was based on knowledge that we see as commonplace. Would we have called Carl Sagan the Christ?

*Reprinted in part from:"Abraham, Gilgamesh, Ishtar, and Biblical History Decoded"

Luke Decoded

*The Gospel of Luke is shown primarily in bold type without analysis, as it is basically the same composition as Matthew without as much detail.

Luke 1

[**11**] And there appeared unto him an **angel** of the **Lord** standing on the **right side** of the **altar** of **incense**.

[26] And in the **sixth month** the angel **Gabriel** was sent from **God** unto a city of **Galilee**, named **Nazareth**,

[27] To a **virgin** espoused to a man whose name was **Joseph**, of the **house** of **David**; and the virgin's name was **Mary**.

[31] And, behold, thou shalt **conceive** in thy womb, and bring forth a **son**, and shalt call his name **JESUS**.

[32] He shall be great, and shall be called the **Son** of the **Highest**: and the Lord God shall give unto him the **throne** of his father **David**:

[33] And he shall reign over the **house** of **Jacob** for ever; and of his **kingdom** there shall be **no end.**

[41] And it came to **pass**, that, when Elisabeth heard the salutation of Mary, the **babe leaped** in her **womb**; and Elisabeth was **filled** with the **Holy Ghost**:

[62] And they made **signs** to his **father**, how he would have him called.

[73] The oath which he sware to our father **Abraham**,

[76] And thou, child, shalt be called the **prophet** of the **Highest**: for thou shalt go before the face of the **Lord** to prepare his ways;

[78] Through the tender mercy of our God; whereby the **dayspring** from on **high** hath visited us,

[79] To give **light** to them that sit in **darkness** and in the **shadow** of **death**, to guide our feet into the way of peace.

[80] And the child **grew**, and **waxed** strong in **spirit**, and was in the **deserts** till the **day** of his shewing unto Israel.

Luke.2

[1] And it came to pass in those days, that there went out a decree from **Caesar Augustus**, that all the **world** should be **taxed**.

[4] And Joseph also went up from **Galilee**, out of the city of **Nazareth**, into Judaea, unto the city of **David**, which is called **Bethlehem**; (because he was of the house and lineage of David:)

[8] And there were in the same country **shepherds** abiding in the field, keeping **watch** over their **flock** by **night**.

[13] And suddenly there was with the **angel** a **multitude** of the **heavenly host** praising God, and saying,

[15] And it came to **pass**, as the **angels** were gone away from them into **heaven**, the shepherds said one to another, Let us now go even unto **Bethlehem**, and **see** this thing which is come to **pass**, which the Lord hath made **known** unto us.

[21] And when **eight days** were accomplished for the **circumcising** of the **child**, his name was called **JESUS**, which was so **named** of the **angel** before he was **conceived** in the **womb**.

[24] And to offer a **sacrifice** according to that which is said in the **law** of the **Lord**, A **pair** of **turtledoves**, or **two young pigeons**.

[34] And **Simeon** blessed them, and said unto Mary his mother, Behold, this **child** is **set** for the **fall** and **rising again** of many in **Israel**; and for a **sign** which shall be spoken against;

[36] And there was one **Anna**, a **prophetess**, the daughter of **Phanuel**, of the tribe of **Aser**: she was of a great age, and had lived with an husband **seven years** from her **virginity**;

[40] And the child **grew**, and **waxed strong** in spirit, filled with **wisdom**: and the **grace** of God was upon

him.

[41] Now his parents went to Jerusalem every year at the **feast** of the **passover**.

[42] And when he was **twelve years old**, they went up to Jerusalem after the **custom of the feast**.

[43] And when they had fulfilled the **days**, as they returned, the child Jesus **tarried** behind in Jerusalem; and **Joseph** and his **mother knew not** of it.

[46] And it came to **pass**, that after **three days** they found him in the **temple**, sitting in the **midst** of the doctors, both hearing them, and asking them questions.

Luke.3

[1] Now in the **fifteenth year** of the reign of **Tiberius Caesar**, **Pontius Pilate** being governor of Judaea, and Herod being **tetrarch** of Galilee, and his brother **Philip** tetrarch of **Ituraea** and of the region of **Trachonitis**, and **Lysanias** the tetrarch of **Abilene**,

[2] **Annas** and **Caiaphas** being the high priests, the word of God came unto **John** the son of **Zacharias** in the **wilderness**.

[4] As it is written in the book of the words of **Esaias** the prophet, saying, The **voice** of one **crying** in the wilderness, **Prepare** ye the **way** of the **Lord**, make his **paths straight**.

[5] Every **valley** shall be **filled**, and every **mountain** and **hill** shall be brought **low**; and the **crooked** shall be made **straight**, and the **rough** ways shall be made **smooth**;

[6] And all **flesh** shall see the **salvation** of **God**.

[14] And the soldiers likewise demanded of him, saying, And what shall we do? And he said unto them, Do violence to no man, neither accuse any falsely; and **be content with your wages**.

[16] John answered, saying unto them all, I indeed baptize you with **water**; but one mightier than I cometh, the latchet of whose shoes I am not worthy to unloose: **he shall baptize you with the Holy Ghost and with fire:**

[17] Whose **fan** is in his **hand**, and he will thoroughly **purge** his floor, and will **gather** the **wheat** into his **garner**; but the **chaff** he will **burn** with **fire unquenchable**.

[22] And the **Holy Ghost descended** in a **bodily shape** like a **dove** upon him, and a voice came from **heaven**, which said, Thou art my beloved **Son**; in thee I am well pleased.

[23] And Jesus himself began to be about **thirty years** of age, being (as was supposed) the son of Joseph, which was the son of **Heli**,

Luke.4

[2] Being **forty days** tempted of the **devil**. And in those days he did eat **nothing**: and when they were ended, he afterward **hungered**.

[13] And when the **devil** had **ended** all the **temptation**, he **departed** from him for a **season**.

[14] And Jesus **returned** in the **power** of the **Spirit** into **Galilee**: and there went out a fame of him through all the **region** round about.

[15] And he **taught** in their **synagogues**, being glorified of all.

[16] And he came to **Nazareth**, where he had been brought up: and, as his **custom** was, he went into the synagogue on the **sabbath day**, and stood up for to read.

[18] The Spirit of the Lord is upon me, because he hath **anointed** me to preach the gospel to the poor; he hath sent me to heal the **brokenhearted**, to preach **deliverance** to the captives, and **recovering** of sight to the **blind**, to set at liberty them that are **bruised**,

[19] To preach the **acceptable year** of the **Lord**.

[23] And he said unto them, Ye will surely say unto me this proverb, **Physician**, heal thyself: whatsoever we have **heard** done in **Capernaum**, do also here in thy country.

[25] But I tell you of a truth, many widows were in Israel in the days of Elias, when the **heaven** was **shut**

up **three years** and **six months**, when great **famine** was **throughout** all the land;

[27] And **many lepers** were in **Israel** in the time of **Eliseus** the prophet; and **none** of them was **cleansed**, saving **Naaman** the Syrian.

[31] And **came down** to **Capernaum**, a city of **Galilee**, and taught them on the **sabbath** days.

[42**] And when it was day**, he **departed** and went into a **desert** place: and the people **sought** him, and came unto him, and **stayed** him, that he should not **depart** from them.

Luke.5

[1] And it **came to pass**, that, as the people **pressed** upon him to hear the **word** of God, he stood by the lake of **Gennesaret**,

[3] And he **entered** into one of the ships, which was **Simon's**, and prayed him that he would **thrust** out a little from the land. And he **sat** down, and taught the people out of the ship.

[6] And when they had this done, they **inclosed** a **great multitude of fishes**: and their net brake.

[13] And he put **forth** his hand, and **touched** him, saying, I will: **be** thou **clean**. And **immediately** the **leprosy departed** from him.

[16] And he **withdrew** himself **into** the **wilderness**, and prayed.

[39] No man also having **drunk old wine** straightway desireth new: for he saith, The old is better.

Luke.6

[1] And it **came** to **pass** on the **second sabbath** after the first, that he went through the **corn fields**; and his disciples plucked the ears of corn, and did eat, rubbing them in their hands.

[5] And he said unto them, That the **Son** of **man** is **Lord** also of the **sabbath**.

[6] And it **came to pass** also on another **sabbath**, that he entered into the **synagogue** and taught: and there was a **man** whose **right hand** was **withered**.

[12] And it **came to pass** in those days, that he went out into a **mountain** to pray, and **continued all night** in prayer to God.

[13] And **when it was day**, he called unto him his disciples: and of them he **chose twelve**, whom also he named **apostles**;

Simon- Peter, Andrew, James, John, Philip, Bartholomew, Matthew, Thomas, James, Simon, and Judas Iscariot.

[17] And he **came down with them**, and **stood** in the **plain**, and the company of his disciples, and a great multitude of **people** out of all **Judaea** and **Jerusalem**, and from the **sea coast** of **Tyre** and **Sidon**, which came to hear him, and to be **healed** of their

diseases;

[18] And **they** that were **vexed** with **unclean** spirits: and they were **healed**.

Luke.7

[1] Now when he had ended all his sayings in the audience of the people, he **entered into Capernaum**.

[5] For he **loveth our nation**, and he hath built **us** a **synagogue**.

[11] And it **came to pass** the **day after**, that he went into a city called **Nain**; and many of his disciples went with him, and much people.

[21] And in that **same hour** he **cured** many of their **infirmities** and **plagues**, and of **evil spirits**; and unto many that were **blind** he gave **sight**.

Luke.8

[2] And certain **women**, which had been healed of **evil spirits** and **infirmities**, **Mary** called **Magdalene**, out of whom went seven devils,

[10] And he said, Unto you it is given to know the mysteries of the kingdom of God: but to others in parables; that seeing they might not see, and hearing they might not understand.

[22] Now it **came to pass** on a certain day, that he went into a ship with his disciples: and he said unto them, Let us **go over** unto the **other side of the lake**.

And they launched forth.

[23] But as they **sailed** he fell **asleep**: and there came **down** a **storm** of **wind** on the lake; and they were **filled** with **water**, and were in jeopardy.

[26] And they arrived at the **country** of the **Gadarenes**, which is over against **Galilee**.

[30] And Jesus asked him, saying, What is thy **name**? And he said, **Legion**: because many **devils** were entered into him.

Luke.9

[22] Saying, The **Son of man** must suffer many things, and be **rejected** of the **elders** and **chief priests** and **scribes**, and be slain, and be **raised** the **third day.**

[34] While he thus spake, there **came** a **cloud**, and **overshadowed** them: and they feared as they **entered** into the **cloud**.

[37] And it **came to pass**, that on the next **day**, when they were **come down** from the **hill**, much people met him.

Luke.12

[**27**] Consider the lilies how they grow: they toil not, they **spin** not; and yet I say unto you, that **Solomon** in all his **glory** was not **arrayed** like one of these.

[38] And if he shall **come** in the **second watch**, or come in the **third watch**, and find them so, blessed are those servants.

[56] Ye hypocrites, ye **can discern** the face of the **sky** and of the **earth**; but how is it that ye do not **discern this time?**

Luke.13

[1] There were present at that **season** some that told him of the **Galilaeans**, whose **blood Pilate** had **mingled** with their **sacrifices**.

[6] He spake also this **parable**; A certain man had a **fig tree** planted in his **vineyard**; and he came and sought **fruit** thereon, and found none.

Luke.17

[24] For as the **lightning**, that **lighteneth** out of the **one part** under **heaven**, **shineth** unto the other part under **heaven**; so shall also the **Son of man** be in his **day**.

[25] But first must he suffer many things, and be **rejected of this generation.**

[29] But the same **day** that **Lot** went out of **Sodom** it **rained fire** and **brimstone** from **heaven**, and destroyed them all.

Luke.20

[25] And he said unto them, **Render** therefore **unto Caesar** the things **which** be **Caesar's**, and **unto God** the **things** which be **God's**.

As long as everyone gets theirs, you can be subdivided to profit the government (33%) and religion (10%)!

Luke.21

[11] And **great earthquakes** shall be in divers places, and **famines**, and **pestilences**; and fearful **sights** and **great signs** shall there be from **heaven**.

[25] And there shall be **signs** in the **sun**, and in the **moon**, and in the **stars**; and upon the **earth distress** of **nations**, with **perplexity**; the **sea** and the **waves** roaring;

Luke.22

[1] Now the feast of **unleavened bread** drew nigh, which is called the **Passover**.

[10] And he said unto them, **Behold**, when ye are **entered** into the **city**, there shall a **man** meet you, **bearing** a **pitcher** of **water; follow** him **into** the **house** where he **entereth** in.

[11] And ye shall say unto the **goodman** of the **house**, The **Master** saith unto thee, Where is the **guestchamber**, where I shall eat the **passover** with my **disciples**?

[19] And he took **bread**, and gave **thanks**, and brake it, and gave unto them, **saying**, This is **my body**

which is given for you: this do in remembrance of me.

[20] Likewise also the **cup** after supper, saying, This cup is the **new testament** in my **blood**, which is shed for you.

[69] Hereafter shall the **Son of man** sit on the **right hand** of the **power** of **God**.

Luke.23

[30] Then shall they begin to say to the **mountains**, **Fall** on **us**; and to the **hills**, **Cover us**.

[44] And it was about the **sixth hour**, and there was a **darkness** over all the **earth** until the **ninth hour.**

[45] And the **sun** was **darkened**, and the veil of the **temple** was rent in the midst.

Luke.24

[7] Saying, The **Son of man** must be delivered into the hands of sinful men, and be **crucified**, and the **third day rise again**.

[18] And the one of them, whose name was **Cleopas,** answering said unto him, Art thou only a **stranger** in **Jerusalem**, and hast not known the things which are **come to pass therein these days?**

The Disaster-Miracle Path of Jesus Christ

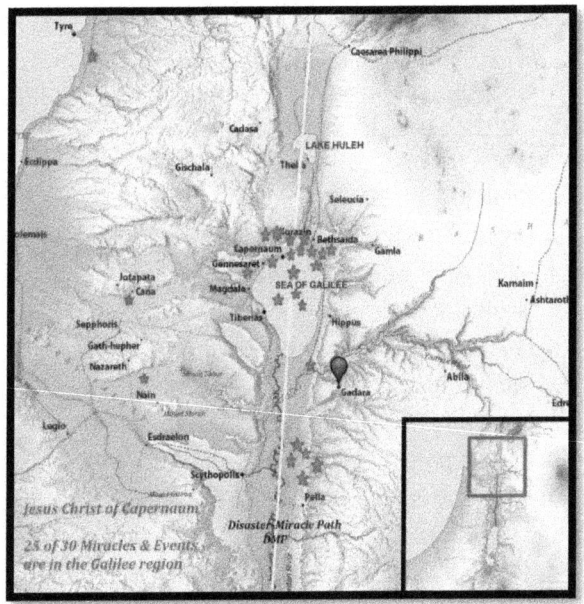

    Of the thirty or so miracles associated with the life of Jesus Christ, an amazing twenty-five events occur in the Capernaum or Sea of Galilee (46 sq. mi.) region. The small area covered by these events are only about 500 square miles, almost half the size of the state of Rhode Island. The earth has almost 396,000 square miles of land, and only 500 square miles were trod to conceive a new religion. The figure of Jesus Christ missed the other 395,500 square miles of the planet, as this concept speaks to a local system of belief. The United States has over a

hundred state parks that dwarf this in protected square miles. Listed below are twenty-five miracles associated with the figure of Jesus Christ, as only five other events tie him to other regions.

<u>Capernaum - Sea of Galilee Miracles</u>
1. Water to wine
2. Heals officials son
3. Heals possessed man
4. Heals Peter's mother in-law
5. Heals sick in the evening*
6. Catches a large number of fish
7. Heals a leper
8. Heals a centurion's servant
9. Heals a paralytic
10. Resurrects a widow's son
11. Calms the sea
12. Heals man possessed of legion
13. Heals internal bleeding
14. Resurrects the daughter of Jairus
15. Heals 2 blind men
16. Heals a mute of demons
17. Feeds 5,000 people*
18. Walks on water*
19. Miraculous multiple healing*
20. Heals a possessed girl
21. Heals deaf man with speech impediment
22. Feeds 4,000 people*
23. Heals a blind man
24. Fish with coin in mouth
25. Catches a great number of fish / post-mortem

Events outside of Capernaum
1. Birth in Bethlehem*
2. Trip to Jerusalem as infant
3. Trip to Egypt
4. Crucifixion

All events occur on a mountain at evening, Capernaum, Sea of Galilee, Nazareth, ships, Gennesaret, Tyre and Sidon Lebanon, Magdala, Jericho, Bethphage, Bethany, and Gethsemane. This region of northern Israel synthesized Judaism, Greek gods, and also Roman gods. This is the reason that Christ is linked to the Temple of Pan in Caesaria Philippi, as Pan was a harvest deity, just as Jesus Christ is as well.

All miracles and events occur from 6:00PM and end at 6:00AM as the Hebrew Watch times for guarding the skies begin at even. The watch times begin with the rise of the Evening Star, Venus, and ends with the rise of the Morning Star, Venus!

Watch Times:
1st Watch: 6:00PM to 9:00PM
2nd Watch: 9:00PM to 12:00AM
3rd Watch: 12:00AM to 3:00AM
4th Watch: 3:00AM to 6:00AM

Jewish Hourly divisions BCE:
1st Hour: 6am to 8am
2nd Hour: 8am to 9am
3rd Hour: 9am to 10am

4th Hour: 10am to 11am
5th Hour: 11am to 12pm
6th Hour: 12pm to 1pm
7th Hour: 1pm to 2pm
8th Hour: 2pm to 3pm
9th Hour: 3pm to 4pm
10th Hour: 4pm to 5pm
11th Hour: 5pm to 6pm
12th Hour: 6pm to sundown

Jesus Christ: Deity of the Solar Harvest

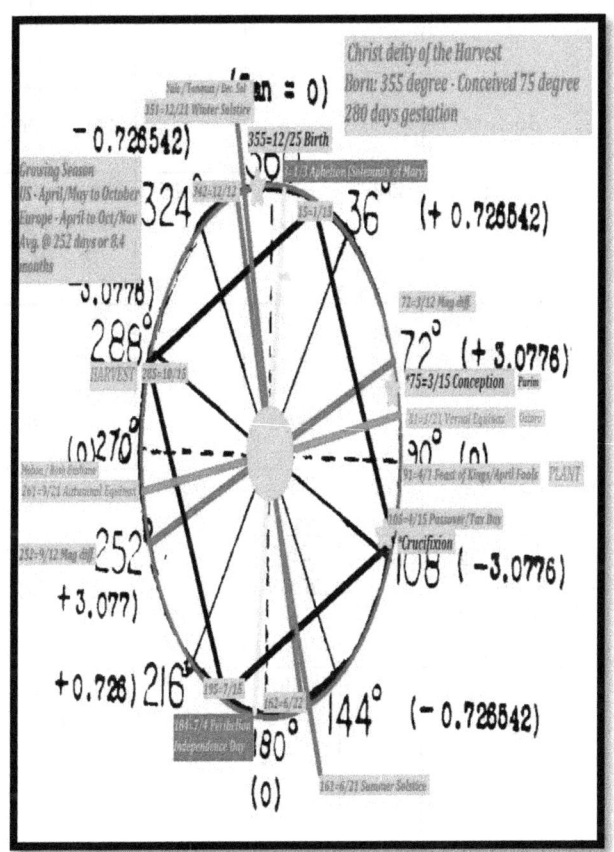

Christ Birth	355°	*0 birth	12/25	Christmas, Tammuz, Horus (Sun/Son is reborn)
Sun in Hell	352°	1-3	12/22	Sun sits at

3 days	to 354°	days	to 12/24	the lowest point annually (dies)
Winter Solstice	351°	4 days	12/21	Shortest day of the year
Harvest season	285°	70 days	10/15	7mo. from planting or 210 days
Autumnal Equinox	261°	94 days	9/21	Onset of Fall
Summer Solstice	161°	194 days	6/21	Longest day of the year
Passover Crucifixion	105°	250 days	4/15	Tax Day
Planting Season & Feast of Kings	91°	264 days	4/1	April Fool's Day
Vernal Equinox	81°	274 days	3/21 Skull & Bones 322 club or 3/22 marks the equinox	Spring begins
Conception of Christ	75°	*280 days before	3/15	Purim

| | | *birth* | | |

The normal human gestational period lasts 280 days or 9 months. If we go from the stated all too familiar birth date of December 25th (351°) backward to the point of conception 9 months prior, we would arrive at March 15th (75°). This shows the clear formulation of Jesus Christ as a harvest deity, as his mother is miraculously impregnated by a god. In order to avoid appearing as a demigod, Jesus Christ is cast as son and father, thus securing his divinity. The 75° point of March 15th also marks Purim on the Jewish calendar, and shows the crafting of this Conception point to align with the Jewish tradition. Purim celebrates the supposed breaching of the walls by the Babylonians on 3/15 around the year 580 BCE. The planting begins in April, as Jesus is always worried about the grapes or the vineyard, just as the demigod Pan. The summer waxes on, as we pass the Summer Solstice on June 21st (161°) the harvest begins to grow healthy. The Autumnal Equinox on September 21st (261°) signals that it is almost time to harvest. By October 15th (285°) most regions have begun to bring in the harvest, in order to finish by November. The harvest is celebrated as Thanksgiving at the 327° point of 11/27, or the third Thursday in November for a bountiful harvest.

December 21st (351°) brings us to the Winter Solstice and the shortest day of the year, as it is said that the Sun dies on the Southern Cross/Crux constellation. The Sun sits at the lowest degree on the horizon from December 22nd (352°) to December 24th (354°), as it descends into hell for 3 days. The Sun/Son is reborn on December 25th (355°), as it begins to rise 1° higher on the horizon signaling warmer weather to come. This process has come to be known as the birth of the Risen Sun/Son.

The Perihelion Miracle of Fatima

The so-called Miracle of the Sun was an event that unfolded on 13 October 1917 which was attended by 30,000 to 100,000 people, who came to Fátima, Portugal. Several newspaper reporters were in attendance and they took testimony from many people who claimed to have witnessed the extraordinary solar activity. According to these reports, the event lasted approximately ten minutes.

The three children who originally claimed to have seen Our Lady of Fátima also reported seeing a panorama of visions, including those of Jesus, Our Lady of Sorrows, Our Lady of Mt. Carmel, and of Saint Joseph blessing the people. The people had gathered because three young shepherd children had predicted that at high noon the "lady" who had appeared to them several times would perform a great miracle in a field near Fátima called Cova da Iria. According to many witnesses, after a period of rain, the dark clouds broke and "the sun" appeared as an opaque, spinning disc in the sky. It was said to be significantly duller than normal, and to cast multicolored lights across the landscape, the shadows on the landscape, the people, and the surrounding clouds. The sun was then reported to have careened towards the earth in a zigzag pattern, frightening those who thought it a sign of the end of the world. Witnesses reported that their previously wet clothes became "suddenly and completely dry, as well as the wet and muddy ground that had been previously soaked because of the rain that had been falling".

Estimates of the numbers present range from 30,000 to 40,000 by Avelino de Almeida, writing for the Portuguese newspaper O Século, to 100,000, estimated by Dr. Joseph Garrett, professor of natural sciences at the University of Coimbra, both of whom were present that day.

The event was attributed by believers to Our Lady of Fátima, a reported apparition of the Blessed Virgin Mary to the children who had made predictions of the event on 13 July 1917, 13 August, and 13 September. The children stated that the Lady had promised them that she would on 13 October reveal her identity to them and provide a miracle "so that all may believe."

Fatima Miracle Analyzed and Explained

The first thing that stood out for me was the prediction that the appearances would take place for 6 consecutive months occurring on the 13th day of each month. The use of "6" and "13" shows the understanding of these mystical numbers used my many religions and secret societies. I have shown previously the frequent occurrences in the use of the number 13 in the United States of America; 13

Colonies, 13th Amendment, Friday the 13th, Chapter 13, 13 Witches in a Coven, 13 sticks in a fasces, just to name a few. The use of the number 13 is also connected to the Star of David. Place a hexagon around the star and assign numerical values from 1 to 12 for each triangle (inner to outward) and 13 for the inner hexagon. The ultimate numerical equivalent breaks down the sum of 6 outer triangles (1-6) = 21, 6 inner triangles (7-12) = 57, and (13) inner hexagon 6 = 13, as 21 + 57 = 78 ÷ 13 = 6 for 666!

The Star of David is an icon that predates the use by the Jewish faithful, and was used as a math teaching tool on one level. I do not know if this symbolism was crafted by those far older than the children who professed this purely esoteric prophecy, as it is apparent that this was indeed crafted from such a system.

This Miracle being totally dependent upon the Sun's appearance on the 13th of each month beginning on April 13th and culminating on October 13th gives away the whole of the event. The 360 day calendar was different than that of the 365 day calendar based on the aphelion and perihelion at that time.

The pre 2344 BCE perihelion (closest to the Sun) of the Earth occurred on October 16th - 10/16 (286°) and does so now on January 3rd - 1/3 (3°).

This shows a 77 day difference in perihelion with adding 5.5 days for a total of 79 days forward movement. The aphelion (furthest from the Sun) of the Earth takes place on April 16th - 4/16 (106°) in pre 2344 BCE, and passes on July 4th - 7/4 - (185°) in 2012 CE. The 6 month time period goes from aphelion on 4/16 to perihelion on 10/16. This 184° point of today versus the 106° then shows a 79° or day difference in the aphelion! The eastern corner of the Great Pyramid is the 195° point and 105° is the pass over point that is celebrated in Christianity, with the Passover being celebrated on the former aphelion point of April the 16th.

Perihelion: 2344 BCE - 10/16   2012 CE - 1/3
Aphelion: 2344 BCE - 4/16*   2012 CE - 7/4
Winter Solstice: 2344 BCE - 12/21
*   2012 CE - 12/21*
Summer Solstice: 2344 BCE - 6/21
*   2012 CE - 6/21*
Vernal Equinox - 2344 BCE - 3/21
*   2012 CE - 3/21*
Autumnal Equinox: 2344 BCE - 9/21
*   2012 CE - 3/21*

* The Great Global Flood occurred on Friday, April 13, 2344 BCE causing the seasons to move forward 74 days, changing the axis of the Earth from 18°s to 23.5°s overnight, and slowed the rotation from 18 hour days to 24 hour days!

The faithful crowd at Fatima waited in the pouring rain for hours peering skyward in anticipation of the miracle unfolding. This belief and hope primed the crowd for a normal event of solar phases being portrayed as a paranormal divine prophecy. The Sun does not dance in the sky, only our gaze does so. With the Sun being at the 2344 BCE perihelion point, an object known as a parhelion or "mock sun", a sundog is a relatively common atmospheric optical phenomenon associated with the reflection/refraction of sunlight by the numerous small ice crystals that make up cirrus clouds or cirrostratus clouds. A sundog is, however, a stationary phenomenon, and one could not explain the reported appearance of the "dancing sun". This parhelion was postulated by author-researcher Joe Nickell, I cannot explain the colors (parhelion/ice) of the Sun or clouds at this point, except to offer that normal atmospheric phenomenon were at work. This shows the Miracle to be solar based lacking any credible claims to the Divine. The search for the Divine often leaves us prey/pray to the mystical aspects of our existence.

Lunar Phases 1917

Date	Degree	Lunar Phase	Lunar Transition
April 13th	103°	Full/3rd Quarter	April 14th
May 13th	133°	Full/3rd	May 14th

		Quarter	
June 13th	163°	3rd Quarter	June 12th
July 13th	193°	3rd Quarter	July 11th
August 13th	223°	3rd Quarter	August 9th
September 13th	253°	3rd Quarter	September 8th
October 13th	283°	3rd Quarter	October 7th

*Solar cycle 15 ran from August 1913 CE to August 1923 CE, with a **solar maximum** of 105.4 sun spots occurring in **August 1917 CE**.*

An annular solar eclipse occurred on December 13, 1917 CE.

In 1917 there were 4 solar eclipses, and 3 lunar eclipses on the records.

*reprinted from: "Holi-days & Religion by Degree & Tangent"

Anti-Masonic Prophecies & Miracles of Our Lady of Good Success

Mariana de Jesus (Mary of Jesus) Torres was born in 1563 in Spain in the province of Viscaya. At age 9 on the day of her first Communion the Lady appeared to her, and told her that she was destined to found an Immaculate Conception Order in the New World. In 1577 when Mariana was only 13 she left with others to Quito, Ecuador.

Six (6) years later at age 19 in 1582, while praying before the Blessed Sacrament she was given a vision of the blasphemy, heresy, and impurity, that would inundate the world in the 20th century. Mother Mariana fell dead during the vision, but was given the choice to return and to suffer the sins of the world.

Six (6) years later at age 25 in 1588, Sister Mariana died a second death on Good Friday, and was subsequently resurrected on Easter Sunday. Sister Mariana died the third and final time on January 16, 1635. Sister Mariana led an extraordinary life, one in which she was imprisoned for 5 years in the convent jail by a group of rogue nuns.

The Lady appeared in Quito, Ecuador, on January 20, 1610 carrying a crozier in her right hand

and the infant Jesus in her left arm. The Lady prophesied that the 20th century would be ruled by Satan almost exclusively by means of the Masonic sects, and that the Devil would try to destroy the sacraments of Confession and Holy Communion. The Conceptionist Sister Mariana also prophesied that; "My hour will arrive when I, in an amazing manner, will overthrow proud Satan, crushing him under my feet, chaining him in the infernal abyss, leaving the church and the land free of the cruel tyranny."

In 1607 Pope Paul V bestowed the name Virgin of Good Success to a statue miraculously found by two Spanish Brothers. These Brothers were of the Order of Minims, and were on their way to Rome seeking Papal confirmation of their Order. A severe storm hits when the two Brothers are passing through Traigueras, as they a dim light in the distance. They find a cave carved like polished stone with fragrant flowers that enshrined a statue of Mary with the infant Jesus in her left arm and a crozier in her right. None of the towns people lay claim to the statue, as it is carried to Pope Paul V who confirms the Order upon hearing of the Miracle and acknowledged the supernatural nature of it. The statue was placed in the Royal Hospital of Madrid and soon became famous for the favors granted through it.

Mother Mariana in a vision with the Lady Mary asked that a statue of her be made as she appears to her there, with the infant Jesus in her right arm, and an Abbess crozier with convent keys in her left hand. The 9th Bishop of Quito, Salvador de Ribera presided over the anointing of the statue on February 2, 1611 (2/2 Imbolc & Ground Hog Day - the 32° - Masonry excels at the 33°) in the Church of the Royal Convent of the Immaculate Conception.

Our Lady of Good Success Analyzed

One must view these Miracles and prophecies under the microscope of that time period. Despite the anti-Masonic prophecies, there is a shared philosophy of the Left Hand Path versus the Right Hand Path. Let us take a look at Sister Mariana Torres first, as she shows the unbelievable structure.

<u>*Sister Maraiana:*</u>

Born 1563 - (Sir Francis Bacon born 1561)

1576 Leaves Spain for Quito Ecuador: Age 13

1st Death 1582 - Age 19 (6 years after arrival)

2nd Death 1588 - Age 25 (6 yeurs after 1st death)

Dies: Good Friday Resurrected: Easter Sunday

3rd Death (final) 1635 (6) - Age 74 (Sir Francis Bacon dies 1626)

The life and deaths of Sister Mariana shows a direct correlation to Masonic or Hermetic systems. At the lucky or unlucky age of 13 she crosses the seas to the New World (Order?), where 6 years later she dies a first death, and then 6 years later she dies a second death, and ultimately dies 49 (Jubilee Passover) years later. The use of 13, 6-6-6, and 49, shows a link to the Masonic and Kabalistic belief systems.

The second death on Good Friday, and subsequent resurrection on Easter Sunday, shows her elevation as a Christ-like assumption of transfiguration. The event and dates show the clearly looted Miracle taken from that of Jesus Christ. I cannot say if he died one death or three, as we have dozens of these occurrences globally on an annual basis. There are countless stories of people waking up in the morgue, undertakers, or even buried alive! Sister Mariana is definitely elevated as she proclaims that she will step on the neck of Satan, and thereby chain him to the abyss.

There seems to have been a dispute in philosophy as to which arm the infant Jesus was supposed to rest, and which hand the crozier that symbolizes ruler-ship should reside. The Left Hand Path is associated with Masonic and Luciferian thought, which is why the statue had to epitomize the Right Hand Path. The Left Hand Path is enacted by the Islamic faithful as they perform the Tawaf while circumambulating the Kabaah, it is kept on the

Left Shoulder in effigy of the zodiac circulating around the heavens. The zodiac sign bindings typically touch month to month with a left foot, hand, horn, etc., to the next monthly sign in a baton passing. This would be considered following the Left Hand Path!

The two Brothers Miraculously finding the statue in their travels, while seeking Papal approval for their Order smells of a fugazy (fake). The timing of these events shows what the Church was trying to rival and mitigate, and it was the alternate system opposing it of Masonry. The life and death of Sir Francis Bacon (1561-1626) almost mirrors that of Sister Mariana (1563-1635), as he was a champion of alternate thoughts and beliefs, and with also being connected to the Rosicrucian Order. These two systems were clashing over new territory and minds in the New World.

The dedication of the statue of Our Lady of Good Success took place on February 2, 1611, and this is an important clue for three reasons.

Reason 1: The King James version Bible is also released in 1611.

Reason 2: The 1611 date is 144,000 (Sealed Revelations) days or 1 Mayan Baktun away from the 20th century and 2012.

Reason 3: February 2nd is the 32nd day of the year, and this day is still celebrated as Imbolc and Ground Hog Day, as an animal is used for divining future events, while we Pass Over to the 33° or day.

The Miracles and interpretations are clearly crafted in my view, as they are related to the climate of the times. I cannot assign validity as the creation is anti one system, but using the same basis of understanding as the other system.

*reprinted from: "Holi-days & Religion by Degree & Tangent"

Conclusions

The peaceful figure that I had come to know as Jesus Christ, has been found to be a compilation of beliefs and deities. All of the disturbing murders, sexual depravity, and religious high roads portrayed in the Bible, Jesus Christ was seemingly the only figure of true love and tolerance. That was until I read the Arabic or Syriac Gospel of the Infancy of the Savior Jesus Christ. The Jesus Christ portrayed in this book is a demigod bent on killing anyone that crossed him improperly.

We often overlook the patently obvious answer, based on the misconception that we already have the answer. The birth of Jesus Christ on December 25th (355°), is remembered in many traditions and religions several thousand years before the birth of Jesus. One of the misconceptions formed in order to validate the belief system, is that events happened in the past to prepare his arrival. The pyramids of Egypt have nothing at all to do with Jesus Christ, as the Egyptian system formed the larger basis for Judaism and Christianity.

The birth of JC is remembered as Horus, Tammuz, Krishna, Dionysus, Apollo, Adonis, Mithras, Yule, and over thirty other solar deities. December 21st (351°) is the Winter Solstice as the shortest day of the year. On December 22nd (352°) the sun is said to die on the Southern Crux/Cross constellation, as it sits at its lowest point annually on the horizon.

The sun begins a descent into hell (below the ecliptic plane), as it descends blow the visible horizon. This descent runs from December 22nd (352°) to the 24th (354°), as the 24th is celebrated as Christmas Eve in tradition.

Christmas Day on the 25th (355°) sees the birth of the new Sun/Son, as it begins the annual ascent of 1° higher being at this degree. The mistaken teachings in regard to solar mechanics, has brought much sorrow and bloodshed to humanity.

When the normal human gestation period of 280 days is accounted for from the December 25th (355°) birth date of Jesus Christ, we arrive at several distinct degrees and tangents. Taking 280° from 355° (December 25th) we get 75° (March 15th)(Purim). This is an all too important seasonal point for planting the annual harvest. The earth reaches the Vernal Equinox on March 21st (81°), as the planting season begins April 1st (91°). The 91st degree is celebrated as, The Feast of Kings, and April Fool's Day. The alignment of these holi-days can only say that a person is an April Fool, for believing in Christianity.

The earth reaches an extremely strong magnetic axis point on March 12th (72°) as an intro to the change in seasons. The magnificent setup to the universe constantly amazes me, as I learn more in the true anatomy to our existence.

The conception of Jesus Christ would have taken place on March 15th (75°) as Mary (earth) is fertilized with the onset of Spring. The harvest typically occurs between October (271°) and November (301°) around the planet. The circle was squared long ago by the Egyptians and other cultures, we must learn to decipher ancient concepts and terminology, while learning who we have to become in the very near future.

Lest there be no mistake as to the true nature of these holi-days, beliefs, degrees, tangents, and seasons:

1° - Jan. 3rd - 1/3 - Aphelion - Solemnity of Mary
72° - March 12th - 3/12 - Solar magnetic point
75° - March 15th - 3/15 - Conception date of Jesus Christ - Purim
81° - March 21st - 3/21 - Vernal Equinox (onset of spring)
91° - April 1st - 4/1 - Planting season begins - Feast of Kings - April Fool's Day
171° - June 21st - 6/21 - Summer Solstice - Litha - June Solstice Day
184° - July 4th - 7/4 - Aphelion - Independence Day
195° - July 15th - 7/15 - 17th of Tammuz
252° - September 12th - 9/12 - Solar magnetic point

261° - September 21st - 9/21 - Autumnal Equinox
271° to 301° - October 1st to November 1st - 10/1 to 11/1 - Harvest Season - Halloween - Samhain - All Saints Day
351° - December 21st - 12/21 - Winter Solstice - Yule - December Solstice Day - Candlemas
352° to 354° - December 22nd to 24th - 12/22 to 12/24 - Annual solar low point
355° - December 25th - 12/25 - Birth of Jesus Christ / Christmas (280 days after conception), Horus, Tammuz, etc.
360° - December 30th - 12/30 - New Years Eve

It is clear that the man known as Jesus Christ has been anthropomorphized as a harvest deity, solar deity, and as the Morning Star Venus. The books of Matthew and Luke show scope of who Jesus Christ was blended with, and the Syriac Gospel shines another ray of light on this subject.

The astrotheology is ever present as there is a constant astronomy lesson. The two year old Christ travels behind his parents as he walks alone This speaks to the sun/son at their backs! The infant Christ comes out of the furnace unscathed, just as Abraham did as well.

Christ wantonly murders two children and a school master, and brings a boy back from death to testify that Jesus had not knocked him off of the rooftop. This clearly shows Jesus as a solar figure, as the people worshipped the sun on their rooftops. Read the whole Syriac /Arabic Gospel to get a much clearer picture. Knowledge is a key that must be used with the proper lock.

www.46stbooks.com

www.ingramcontent.com/pod-product-compliance
Lightning Source LLC
Chambersburg PA
CBHW071703040426
42446CB00011B/1895